D0075089

THE FOUNDATIONS OF NEUTRON TRANSPORT THEORY

American Nuclear Society
and
U.S. Atomic Energy Commission

MONOGRAPH SERIES ON NUCLEAR SCIENCE AND TECHNOLOGY

ALLEN G. GRAY
Series Editor
American Society for Metals

JOHN H. GRAHAM
Series Editor
American Nuclear Society

The Foundations of Neutron Transport Theory
RICHARD K. OSBORN and SIDNEY YIP

Coolant Chemical Technology of Aqueous Heterogeneous Reactor Systems
PAUL COHEN

Alkali Metal Handling and Systems Operating Techniques
J. W. MAUSTELLER, F. TEPPER and S. J. RODGERS

Ceramic Fuel Elements
ROBERT B. HOLDEN

Non-Destructive Fuel Assay
WARREN J. McGONNAGLE

Fabrication of Refractory Metals
JAMES F. SCHUMAR and ROSS MAYFIELD

Irradiation Behavior of Nuclear Fuels
J. A. L. ROBERTSON

Fuel Elements in Operational Nuclear Power Reactors
MASSOUD T. SIMNAD

Liquid Metal Heat Transfer
O. E. DWYER

Advanced Metalworking Processes
E. E. BISHOP and F. L. ORRELL

Dispersion Type Fuel Elements
A. N. HOLDEN

THE FOUNDATIONS OF NEUTRON TRANSPORT THEORY

RICHARD K. OSBORN
College of Engineering, University of Michigan
Ann Arbor, Michigan

SIDNEY YIP
Massachusetts Institute of Technology
Cambridge, Massachusetts

Prepared under the auspices of
the Division of Technical Information
United States Atomic Energy Commission

GORDON AND BREACH, SCIENCE PUBLISHERS, INC.
NEW YORK · LONDON · PARIS

150 Fifth Avenue, New York, N.Y. 10011

Library of Congress Catalog Card Number 66-24007

Distributed in the United Kingdom by:

Blackie & Son, Ltd.
5 Fitzhardinge Street, London W.1, England

Distributed in France by:

Dunod Editeur
92, rue Bonaparte, Paris 6, France

Distributed in Canada by:

The Ryerson Press
299 Queen Street West, Toronto 2B, Ontario

Preface

There are at least three reasons why the authors felt that a monograph such as this might prove useful. For the past fifteen years or so there have appeared many texts and treatises which have presented extensive studies at all levels of sophistication of the solutions of the neutron transport equation. However, the origins and limitations of this equation have been given little or no attention. But the fission reactor technology (like the fusion technology and many other areas of modern engineering) is demanding a deepening awareness of the subtle relationship between microscopic cause and macroscopic effect. Thus we felt that an initiation of an exploration into the foundation of the neutron transport equation was a needed complement to the examination of its solutions.

The subject matter summarized in this monograph was initially generated in bits and pieces within the context of various courses offered to the nuclear engineering students at the University of Michigan. Thus a second reason for the preparation of this material in its present form was to provide an integrated treatment of an integral topic. For example, it is quite conventional to separate the discussion of the transport equation from the study of microscopic reaction rates. This is both natural and necessary from the pedagogical point of view, particularly at the introductory level. Nevertheless it seems important that at some point the essential unity of these concepts be restored, and this unity manifests itself in the study of the origins of the transport equation—not its solutions.

Thirdly, it is probably inevitable that the analytical tools available to the engineer at any given instant in time will eventually become inadequate to his tasks. Indeed this may be the case in the reactor technology today with respect to the matter of interpreting neutron fluctuation measurements. Thus a potentially practical purpose may be served by this work in that it suggests a pathway along which generalization of the usual description of the reactor may be sought deductively rather than inductively.

This book is not intended to be a text book, nor is it aimed at particular areas of specialization. It deals with a small, well-defined topic, which, however, has broad implications. It is thus anticipated that graduate students, teachers, and research workers in nuclear engineering, physics, and chemistry (many of the principles and techniques of analysis carry over intact from a study of neutron transport to the study of the kinetic theory of reacting gases) might find herein something of interest to them. We have used whatever mathematical tools and physical notions we have found necessary or convenient—usually without providing any background information. Nevertheless we have attempted to present the argument in a sufficiently self-contained way that the bulk of the discussion can be followed without too much reference to background material.

No attempt has been made to compile a comprehensive bibliography. In fact the referencing is admittedly spotty, casual and enormously incomplete. However some care has been taken to see to it that points of connection between the topic discussed here and related topics discussed elsewhere are referenced for the reader's general interest. Also some forethought was exercised to supply references which in themselves provide good bibliographies.

The authors are grateful to Professor George Summerfield for his careful reading of the manuscript and his helpful comments and criticisms, to Professor Ziya Akcasu for his assistance with the perturbation method used here for the calculation of nuclear reaction rates, and to Mr. Malcolm Ferrier whose interest in and encouragement of this work was crucial to its fruition. One of us (S.Y.) gratefully acknowledges the University of Michigan Institute of Science and Technology for a postdoctoral fellowship and the Michigan Memorial-Phoenix Laboratory for hospitality during the course of this work.

Table of Contents

List of Symbols

In the following some of the more frequently used mathematical symbols are defined. Whenever possible the equation in which the symbol is first introduced is given and the reader is referred to the text for definition. Some symbols are used for more than one quantity, but their meanings should be clear from the context.

A	— Atomic mass
$A(\mathbf{K}'k'; \mathbf{K}k)$	— Scattering "frequency", Eq. 5.2
a	— "Free-atom" scattering length
$a^{+}(\mathbf{X}, \mathbf{K}, s)$, $a(\mathbf{X}, \mathbf{K}, s)$	— Creation, destruction operators for neutrons at $(\mathbf{X}, \mathbf{K}, s)$
$a_1, a_2, a_{\pm}, a_{\text{coh}}, a_{\text{inc}}$	— Scattering lengths, see Eqs. 4.71, 4.72, and 4.76
B^{A+1}	— $(m + M_A - M_{A+1})\, c^2$
$B_\alpha^J(\mathbf{P}', \mathbf{P})$, $B_{\alpha\sigma}^J(\mathbf{P}'', \mathbf{P}, \mathbf{P}')$	— Fission frequencies, Eqs. 2.80 and 5.43b
D	— Density operator, Eq. 2.29; see also Eq. 4.45
$D_{mn}(t)$	— Density matrix, Eq. 2.28
E, E_K	— Neutron energies, $mv^2/2$ or $\hbar^2 K^2/2m$
$E(\mathbf{X}, \mathbf{x})$	— Step function, Eq. 2.2
E_R	— Nuclear recoil energy, Eq. 4.33
$E_\varkappa$	— Photon energy, $\hbar c \varkappa$
E_{Kk}, E_r	— Relative energies, Eq. 4.4
E_k^A	— External energy of mass A nucleus in external state k
E_α^A	— Internal energy of mass A nucleus in internal state α
E_α^{A+1}	— Excitation energy of αth level in mass $(A + 1)$ compound nucleus
E_α^{A+1*}	— $E_\alpha^{A+1} - B^{A+1}$

$\mathscr{E}_\alpha$	— $E_\alpha^{A+1} + s_\alpha + B^{A+1}$
$F(\mathbf{X}, \mathbf{K}, t)$ or $F_1^{(n)}(\mathbf{X}, \mathbf{K}, t)$	— Coarse-grained neutron singlet density, Eq. 2.27
$F_s(\mathbf{X}, \mathbf{K}, t)$	— Spin-dependent coarse-grained neutron density, Eq. 3.65
$F_\lambda(\mathbf{X}, \varkappa, t)$	— Coarse-grained photon density, Eq. 3.67
$F_1^{(\alpha)}(\mathbf{X}, \mathbf{K}, t)$	— Coarse-grained singlet density for α particles, Eq. 5.44a
$F_2(\mathbf{X}, \mathbf{K}, \mathbf{X}', \mathbf{K}', t)$ or $F_2^{(n)}(\mathbf{X}, \mathbf{K}, \mathbf{X}', \mathbf{K}', t)$	— Coarse-grained neutron doublet density, Eq. 2.70
$F_2^{(\alpha)}(\mathbf{X}, \mathbf{K}, \mathbf{X}', \mathbf{K}', t)$	— Coarse-grained doublet density for α particles, Eq. 5.44b
$F_2^{(\alpha n)}(\mathbf{X}, \mathbf{K}, \mathbf{X}', \mathbf{K}', t)$	— Coarse-grained cross density for neutrons and α particles, Eq. 5.44c
$f(\mathbf{X}, \mathbf{K}, t)$	— Neutron density, Eq. 2.30
$f_A(\mathbf{X}, \mathbf{k}, t)$	— Density of mass A nucleus, Eq. 4.27
$\mathscr{F}(\mathbf{P}' \to \mathbf{P})$	— Scattering frequency, Eq. 2.80
$\mathscr{F}^D(\mathbf{P} \to \mathbf{P}')$	— Detection frequency, Eq. 5.45
$G(z)$	— Resolvent operator, Eq. 3.2
$G(\mathbf{r}, t)$	— See Eq. 4.78
H_s	— Hamiltonian of system which interacts with the neutrons
H'	— See Eq. 2.42
$\hbar$	— $h/2\pi$, where h is Planck's constant
$I_n(x)$	— Modified Bessel function of order n and argument x
$\mathbf{K}$	— Neutron wave vector, $\mathbf{P}/\hbar$; also α particle wave vectors
$\{K_l\}_J$	— A set of J neutron wave vectors
$\{K\}_{JK}$ $(\{K\}'_{JK})$	— A set of J wave vectors which contains (does not contain) the wave vector $\mathbf{K}$
$\{Ks\}_J$	— Wave vector and spin labels of the J neutrons produced by fission
k	— Quantum label of nuclear external state, Eq. 3.38

$\mathbf{k}$	— Wave vector of a nucleus, $\mathbf{p}/\hbar$
k_B	— Boltzmann's constant
L	— Cell length in coarse-grained configuration space
l	— Superscript or subscript denotes a particular nucleus
$\mathscr{L}(\mathbf{v}' \to \mathbf{v})$	— Fission frequency, Eq. 3.99
M	— Mass of the nucleus
m	— Neutron mass
m_L	— Mass of Lth nucleon
$N(\mathbf{X}, \mathbf{K}, s)$, $N'(\mathbf{X}, \mathbf{K}, s)$	— Neutron occupation number at $(\mathbf{X}, \mathbf{K}, s)$ in the state $\|n\rangle$, $\|n'\rangle$
$N_A(\mathbf{X})$	— Number of mass A nuclei in the spatial cell centered at $\mathbf{X}$
n	— Nuclear density
$\|n\rangle$	— Neutron state, Eq. 2.18; system state, Eq. 3.35
$\mathscr{N}$	— Total number operator
P	— See Eq. 4.45
$\mathbf{P}$	— Neutron momentum; momentum of α particle
P_k, $P(\mathbf{k})$, $D_{kk}(t)$	— Distribution of external nuclear states
$\mathbf{p}_l^L$	— Nucleonic momentum conjugate to $\mathbf{r}_l^L$
$\mathbf{Q}$	— Neutron momentum transfer divided by $\hbar$
$R_T(\mathbf{X}, \mathbf{K})$	— Total reaction rate
$R_s(\mathbf{X}, \mathbf{K})$, $R_s(\mathbf{X}, \mathbf{K}' \to \mathbf{K})$	— Scattering reaction rate, see Eq. 3.95 and 3.97
$\mathbf{R}_l$	— Position vector of center-of-mass of lth nucleus
$R_{nn'}$	— Reduced transition matrix, Eq. 3.33
$r_{nn'}$	— Reduced reaction matrix with neutron and photon number dependence extracted, Eq. 3.51
$\mathbf{r}_l^L$	— Position of Lth nucleon in lth nucleus

$R^c_{n'n}$, $R^c_{k'\varkappa'\lambda',k\mathbf{K}s}$, $r^c_{k'\varkappa'\lambda',k\mathbf{K}s}$, $R_c(\mathbf{X}, \mathbf{K})$ — See Eqs. 3.42, 3.51, 3.52, and 3.94

$R^F_{k_1\alpha_1,k\mathbf{K}s}(\{Ks\}_J)$, $r^F_{k_1\alpha_1,k\mathbf{K}s}(\{Ks\}_J)$
$R_F(\mathbf{X}, \mathbf{K})$, $R_F(\mathbf{X}, \mathbf{K}' \to \mathbf{K})$ — See Eqs. 3.81, 3.82, 3.96, and 3.98

$R^{Es}_{k'\mathbf{K}'s',k\mathbf{K}s}$, $r^{Es}_{k'\mathbf{K}'s',k\mathbf{K}s}$ — See Eqs. 3.68 and 3.69

$R^{Is}_{k'\mathbf{K}'s'\varkappa'\lambda',k\mathbf{K}s}$, $r^{Is}_{k'\mathbf{K}'s'\varkappa'\lambda',k\mathbf{K}s}$ — See Eqs. 3.87 and 3.88

s — Neutron spin orientation label

s_α — Shift function, Eq. 3.49

U_l — Reduced potential for elastic potential scattering by lth nucleus, Eq. 3.73

$\bar{U}_l$ — See Eq. 4.19

$U^{Nl}_{\alpha''0}(U^{Nl}_{0\alpha''})$ — Nuclear matrix element for emission (absorption) of a neutron by lth nucleus

$U^{Rl}_{0\alpha''}(U^{Rl}_{\alpha''0})$ — Nuclear matrix element for emission (absorption) of a photon by lth nucleus, Eq. 3.55

V — $V^N + V^\gamma$

$\mathbf{V}$ — Nuclear velocity

V^N — Neutron-nuclear interaction

V^γ — Photon-nuclear interaction

$\mathbf{v}$ — Neutron velocity

$v_0(r)$, $v_1(r)$ — Potential functions for elastic potential scattering, Eq. 3.71

$W_{nn'}(\tau)$ — Transition probability per unit time, Eq. 2.56

$W^c_{n'n}$, $W^{sG}_{n'n}$, $W^{sL}_{n'n}$, $W^F_{n'n}$ — See Eq. 3.39

$\bar{w}^c_{\mathbf{K}}$, $\bar{w}^s_{\mathbf{K}\to\mathbf{K}'}$, $w_s(\mathbf{P} \to \mathbf{P}')$, $\bar{w}^F_{\mathbf{K}\to\{\mathbf{K}_l\}_J}$ — See Eqs. 2.65, 2.67, 2.73, and 2.74

$\mathbf{X}$ — Position vector locating the center of a spatial cell

$\mathbf{x}_l$ — Equilibrium position vector of the lth nucleus in a crystal

Z — See Eq. 4.45; see also Eq. 5.23

Greek Letters

α — Quantum label of nuclear internal state

β — $(k_BT)^{-1}$

Γ_α — Width function, Eq. 3.50

$\Gamma_\alpha^{(R)}$ ($\Gamma_\alpha^{(N)}$)	— Partial radiation (neutron) width for αth level
$\gamma_n(z)$	— Width and shift function
ΔE	— $E_K - E_{K'}$
$\delta(\mathbf{x} - \mathbf{x}')$	— Dirac delta
$\delta_{\mathbf{xx}'}$, $\delta_K(\mathbf{x} - \mathbf{x}')$	— Kronecker delta
η	— See Eq. 4.45
θ	— Scattering angle (laboratory)
$\boldsymbol{\kappa}$	— Photon wave vector
λ	— Photon polarization; also as effective range in neutron-nuclear interaction
λ_0, λ_1	— See Eq. 3.72
$\lambda\!\!\!^-$	— Reciprocal neutron wave vector, K^{-1}
μ	— Reduced mass; also as chemical potential, Eq. 5.6
$\varrho_1(\mathbf{X}, \mathbf{K})$	— Neutron number operator, Eq. 2.16
Σ_c, Σ_D, Σ_F, Σ_s, Σ_t	— Macroscopic cross section for capture, detection, fission, scattering and total reaction
σ	— Microscopic cross section
$\sigma(E \to E', \theta)$	— Energy and angle differential scattering cross section, Eq. 4.36
σ_s^p, σ_s^r, σ_s^i	— Microscopic cross section for potential, resonant, and interference scatterings
$\varphi(\mathbf{X}, \mathbf{K}, \mathbf{x})$	— Cell function, Eq. 2.1
χ	— See Eq. 4.25
ω	— Oscillator frequency
ω_n, $\omega_{nn'}$	— See Eq. 3.17

I

Introduction

Theoretical studies of neutron distributions are usually based upon the transport equation

$$\left(\frac{\partial}{\partial t} + \mathbf{v}\cdot\nabla + v\Sigma_t\right) f(\mathbf{x}, \mathbf{v}, t)$$
$$= S(\mathbf{x}, \mathbf{v}, t) + \int d^3v'\, v'\, \Sigma_s(\mathbf{x}, \mathbf{v}')\, \mathscr{F}(\mathbf{v}' \to \mathbf{v}) f(\mathbf{x}, \mathbf{v}', t) \qquad (1.1)$$

Here Σ is a macroscopic cross section, $\mathscr{F}$ is a scattering frequency, and S is a neutron source which may or may not depend upon f. The distribution function $f(\mathbf{x}, \mathbf{v}, t)$, is the neutron singlet density in phase space or the expected number of neutrons per unit phase volume to be found at the phase point $(\mathbf{x}, \mathbf{v})$ at time t.

The motivation for initiating a critical analysis of the neutron transport equation does not arise from the failure of this equation in practical applications. Quite the contrary, in fact, the usefulness of Eq. 1.1 in an overwhelmingly large class of problems of interest in reactor technology has certainly been established beyond doubt. The complete acceptance of this equation, however, has resulted in an atmosphere in which little consideration has been given to the exploration of its foundations.* This is unfortunate in our opinion for, as the utility of Eq. 1.1 increases, our need to understand its origin and validity is also enhanced.

The conventional method of deriving the neutron transport equation is based mainly on plausibility arguments.[2] From a purely theoretical point of view it is unsatisfactory for at least two reasons. The approach is phenomenological in that the result is not derived from a more fundamental description. Secondly, the various cross sections are introduced

---

* A summary of the basic assumptions of neutron transport theory has been given by Wigner.[1]

empirically, hence the treatment must rely upon other more quantitative theories for their calculation. Consequently it seems desirable to turn to a microscopic theory to see to what extent this well-known and highly useful description can be justified from a more fundamental starting point.

Because the neutron transport equation can be considered as a special linear variant of the Boltzmann equation, one may look for a parallel development in the kinetic theory of gases where the latter equation plays a similarly dominant role. It is to be observed that the Boltzmann equation, which was also first obtained on an intuitive basis, had been in use some 32 years before attempts at systematic derivations based on the Liouville equation were made.[3] For the purpose of studying dilute systems, neither the structure nor the physical content of the original equation has been altered by the later, more rigorous investigations. On the other hand, these derivations have contributed significantly to our detailed understanding of the validity of this famous equation. Moreover, the deductive approach has yielded a logical basis for developing transport theories for dense gases and liquids, an area of research that is currently actively being pursued.

To a certain extent, similar advances can be achieved by a systematic study of the basic equation in the theory of neutron transport. Even if we are only able to duplicate the results of the conventional treatment, the analytical approach should yield a set of sufficient conditions for the applicability of Eq. 1.1. Since the analysis would then begin at the level of "first principles", the description can be made essentially self-contained in the sense that both the method used for deriving the transport equation and that used for evaluating the relevant cross sections can be developed from the same basis. Finally, this approach should also be suitable for studying the higher-order neutron densities that would arise in problems concerning fluctuations and correlations. In the present work an attempt is made to realize these anticipations.

It is well known that an axiomatic basis for the theoretical examination of macroscopic systems is provided by the Liouville equation. Although there is a deceptively classical appearance to Eq. 1.1, it turns out to be practically necessary for us to start with the quantum Liouville equation. This necessity manifests itself almost as soon as the problem is posed, and in several different ways. In the first place, such an approach requires some notion of a Hamiltonian for the system and, though such a notion is at best fuzzy and incomplete in the present problem, it is almost unthinkable classically. That is to say, there does not seem to be any classical formalism suitable for treating systems in which particles of a given kind are not conserved. On the other hand, quantum field theory provides a convenient and powerful method of describing the destruction and creation of particles as a result of interactions.

There is a more fundamental reason for resorting to quantum analysis in developing a more quantitative and unified theory of neutron balance. This arises because neutron-nuclear interactions are truly quantum phenomena; in particular the existence of resonance implies that the discretness of nuclear energy states will have an explicit influence on the behavior of the neutron distributions. Again it is difficult, and inappropriate in our opinion, to treat this aspect of the problem in classical terms, particularly since quantum-mechanical calculations exist that allow us to take into account not only the energy states of the individual nucleus but also those of the macroscopic medium (e.g. phonon states in a crystal).

Another reason for the quantum treatment lies in the proper interpretation of an observable density in phase space. In the absence of further qualifying comments, the density described by Eq. 1.1 is ambiguous. As already mentioned, $f(\mathbf{x}, \mathbf{v}, t)\, d^3x\, d^3v$ represents the expected number of neutrons to be found in the phase volume $d^3x\, d^3v$ about the phase point $(\mathbf{x}, \mathbf{v})$ at time t. In view of the uncertainty principle this interpretation becomes meaningless, if the volume element $d^3x\, d^3v$ is taken in a limiting sense, and if not in a limiting sense then how? In a classical treatment this issue does not arise, but one encounters other difficulties in interpretation.* Rather than giving up the physical meaning of the distribution function, we find it necessary, if only as a matter of principle, to give a more concise definition of the particle density of interest. As we shall demonstrate, it is possible in a quantum formulation to answer this question operationally and unambiguously, though not necessarily uniquely. This follows because quantum field theory provides us quite naturally with an operator representative for the number of particles of a given type in a phase cell of sufficient volume.

There are other peculiarly quantum effects which in principle will modify the structure of the transport equation in its final form. While

* See Grad, reference 3, p. 218.

these details are not expected to be of much practical significance, it seems appropriate that in a systematic study their existence should at least be recognized. For example, the velocity distribution of fully thermalized neutrons, strictly speaking, should be of the Fermi-Dirac type rather than the well-known Maxwellian. Admittedly it is difficult to visualize situations in which the two distributions are distinguishable; nevertheless, this property of the neutron distribution can be shown to be a consequence of the derived transport equation.

In concluding these introductory remarks we emphasize that the primary aim of the following work is to provide a more logical foundation of neutron transport theory in its present state of development. Our efforts will be devoted almost exclusively to the justification of the transport equation within the framework of current analytical methods. It will be seen that we are only partially successful in this attempt, because a number of approximations are required in the derivation, and these will be merely stated but not analyzed and evaluated. Evidently if we are to understand the quantitative validity of Eq. 1.1, these approximations must be thoroughly studied. The approximations in question are exceedingly difficult to clarify, and we do not pretend that we either understand all their implications or are even able to carry out the necessary analysis required to attempt this understanding. That the theoretical defense of the mathematical description represented by Eq. 1.1, which may appear quite self-evident on intuitive grounds, requires such involved considerations is an indication that detailed understanding of neutron transport at the microscopic level is still lacking. Consequently there is much to be said for the attempt to delineate the specific areas of difficulty in operational terms, even if the effort to evaluate the approximations has yet to be expended. In this sense the present work, which consists mainly of putting known results into new perspective, represents only the first portion of the ultimate solution. It is clear that to complete this solution further investigations, highly specialized and even more involved, are required. Although no truly new results, except for the generalization to the study of higher-order densities, are obtained, the present approach brings many aspects of the neutron problem into contact with other microscopic transport theories. This identification, which has not been emphasized previously, is potentially quite useful since one can then apply to the study of neutron transport the powerful methods of analysis extensively developed in recent studies of irreversible processes.

The work is presented in four parts. In Chapter II a kinetic equation having the same physical content as that of Eq. 1.1 is derived in coarse-grained phase space. The coarse-graining procedure used allows us to define an appropriate neutron singlet density, which avoids the difficulty in interpretation mentioned earlier. The time evolution of this density is studied under the assumption that there exists a time scale long compared to neutron-nuclear interaction times but short compared to some average time between interactions.* That an approximation of this type seems necessary is quite evident in view of the fact that the desired transport equation treats transport (free flight) and collision processes separately. In the present approach these two processes are moreover treated with different approximations. The conventional description of transport, characterized by the term $\mathbf{v} \cdot \nabla f$, is explicitly obtained and, as one may expect in a coarse-grained analysis, this is only an approximate result. To see this one merely has to note that particles of a given momentum require a volume for localization of least linear dimension greater than the associated de Broglie wavelengths. But the difference between the distribution function on opposite sides of this volume cannot generally be represented solely in terms of the first derivative. Collision effects are discussed only very schematically, primarily to exhibit the resulting equation in a form that is readily identifiable with Eq. 1.1. In fact, when quantum effects and the discrete nature of phase space are ignored, and when finite differences are interpreted as derivatives, the conventional neutron transport equation emerges as the desired result of this study.

Detailed considerations of binary neutron-nuclear interactions are discussed in Chapters III and IV. A brief development of damping theory is first given. This theory enables us to calculate transition probabilities for direct and resonance reactions taking into account both nuclear and macroscopic medium effects. Since we do not intend to present a detailed theory of neutron-induced reactions, we consider only the most relevant reactions from a general standpoint, namely radiative capture, elastic and inelastic scattering, and fission. Moreover, our attention to details must necessarily decrease with increasing complexity of the interaction. As the discussion must touch upon nuclear

* The existence of widely different time scales is an important assumption basic to the study of irreversible processes in gases at low density.[3] For example, Mori and co-workers[4] have emphasized that this procedure is equivalent to Kirkwood's concept of time-smoothing.[5]

forces (which are less than fully understood at the present time), our treatment must be in some respects implicit rather than explicit. However, for the purpose of obtaining useful estimates of reaction rates, a number of well-known results can be deduced with a minimum knowledge of details of nuclear forces. Appropriate formulas for the transition probabilities are derived in Chapter III. These formulas are then examined for some of the salient features of the specifically nuclear aspects of these reactions.

To complete the discussion of neutron-nuclear collisions, the influence of the macroscopic medium on the reactions is studied in Chapter IV. Here only radiative capture and elastic scattering (both potential and resonance) by nuclei in gases and crystals are considered. The emphasis is on idealized systems because exact calculations are then feasible. The results obtained also provide the usual starting point for deriving approximate but more useful expressions for these cross sections.

Finally, in Chapter V two disparate and specialized aspects of neutron balance are discussed. The first has to do with the nature of the velocity distribution of thermal neutrons, while the second relates to the study of higher-order densities. The discussion of thermodynamic distributions is conventional to the theory of gas dynamics, and it is included here because the results are frequently referred to in the neutron context. It is not a discussion of neutron thermalization, but only of some of the concepts that underlie what might be called neutron thermodynamics. The second topic, however, is unconventional in that it represents a systematic development of the theory of second- (and, by implicit extrapolation) higher-order distribution functions in systems in which particles of various kinds are created and destroyed. These distribution functions are of direct interest in interpreting fluctuation and correlation experiments on multiplying systems. Since the implications of the present theory are still currently being explored, we shall restrict our attention to the development of general equations and only some brief remarks concerning applications.

References

1. Eugene P. Wigner, Proceedings of the Eleventh Symposium in Applied Mathematics, American Mathematical Society, Providence, R.I., 1961, p. 89.
2. A. M. Weinberg and E. P. Wigner, *The Physical Theory of Neutron Chain Reac-*

tors, University of Chicago Press, Chicago, Ill., 1958; R. V. Meghreblian and D. K. Holmes, *Reactor Analysis*, McGraw-Hill Book Company, Inc., New York, 1960; B. Davison, *Neutron Transport Theory*, Oxford University Press, London, 1957.

3. H. Grad, Handbuch der Physik, XII, 205 (1958). The literature on the Boltzmann equation is enormous. Grad's work is cited because it is one of the most comprehensive critical studies of the Boltzmann equation from the point of view of classical mechanics. Here the important work of Born and Green, Kirkwood, Bogoliubov, and others are also mentioned. Interested reader should also see E. G. D. Cohen in *Fundamental Problems in Statistical Mechanics*, North-Holland, Amsterdam, 1962, and Mori *et al.*, reference 4.
4. H. Mori, I. Oppenheim and J. Ross in *Studies in Statistical Mechanics*, North-Holland, Amsterdam, 1962, edited by J. deBoer and G. E. Uhlenbeck, Volume 1.
5. J. G. Kirkwood, *J. Chem. Phys.* **14**: 180 (1946); **15**: 72 (1947).

II

A Transport Equation in Coarse-Grained Phase Space

It was pointed out in the introductory remarks that the conventional neutron density $f(\mathbf{x}, \mathbf{v}, t)$ cannot be interpreted as an observable quantum-mechanical entity because of the uncertainty principle. To avoid this difficulty, we shall at the outset introduce a discrete phase space. Such a space can be generated by dividing the continuum into cells and then representing all points in each cell by the coordinate of its center. Particle densities are then defined in terms of these coarse-grained coordinates, and no attempt is to be made to determine the location of a particular particle within any given cell.

In a multiplying and/or an absorbing system, the number of neutrons in a given region in phase space is constantly changing, not only due to the natural flow of these particles but also due to fission and absorption processes. The creation and destruction of neutrons can be quite conveniently described in the formalism of second quantization by representing the particles by a two-component spinor field operator $\psi_j(\mathbf{x})$, where $j = 1$ or 2 is the spinor index. The field formalism plus a procedure for coarse-graining phase space enables us to obtain a particular representation of the number operator, the eigenvalue of which gives the number of particles in a given region in phase space. In terms of the number operator, a coarse-grained quantum-mechanical analogue of $f(\mathbf{x}, \mathbf{v}, t)$ can be defined. This new neutron density will be the quantity for which a transport equation is deduced, and thus it provides the basis of the present investigation of neutron transport theory.

A. Some Basic Formalism

We will first review some of the fundamental concepts and introduce the notations that will allow us to define a coarse-grained particle density in the next section. To introduce the coarse-grained phase space in operational terms we divide the configuration space into identical cubical cells with edge length L. Let an arbitrary point in configuration space be denoted as $\mathbf{x}$ and let the set of position vectors $\{\mathbf{X}\}$ denote the cell centers. The coarse-graining procedure now consists of introducing the cell function,[1]

$$\varphi(\mathbf{X}, \mathbf{K}, \mathbf{x}) = L^{-3/2} E(\mathbf{X}, \mathbf{x}) \, e^{i\mathbf{K} \cdot \mathbf{x}} \tag{2.1}$$

where

$$E(\mathbf{X}, \mathbf{x}) = \prod_{i=1}^{3} E(X_i, x_i) \tag{2.2}$$

$$\begin{aligned} E(X_i, x_i) &= 1 \qquad \text{when} \qquad X_i - L/2 < x_i < X_i + L/2 \\ &= 0 \qquad \text{otherwise} \end{aligned}$$

The cell function $\varphi(\mathbf{X}, \mathbf{K}, \mathbf{x})$ is seen to describe a plane wave which is nonvanishing only within the cell centered at $\mathbf{X}$. In Eq. 2.2 the step function, $E(\mathbf{X}, \mathbf{x})$, is not defined at the end points. However, if it is represented by the integral

$$E(X_i, x_i) = \int_{X_i - L/2}^{X_i + L/2} \delta(x_i - y) \, dy \tag{2.3}$$

it can readily be shown that

$$E(X_i, X_i \pm L/2) = \tfrac{1}{2} \tag{2.4}$$

$$\frac{\partial E}{\partial x_i}(X_i, x_i) = \delta(x_i - X_i + L/2) - \delta(x_i - X_i - L/2) \tag{2.5}$$

These relations describe the behavior of the cell function at the boundaries, and will be used in the description of particle transport. For mathematical convenience we will apply periodic boundary conditions at the interfaces and thereby restrict the components of the wave vector to take on discrete values, $K_i = 2\pi M_i/L$, where M_i is any positive or negative integer or zero. Hence the decomposition of configuration space results in a transformation of the continuous momentum space to a lattice of discrete points.

The coordinates of $\mathbf{X}$ and $\hbar\mathbf{K}$ are to be regarded as coarse-grained variables in our description of particle densities. The phase point $(\mathbf{X}, \hbar\mathbf{K})$

is seen to represent a cubical region of volume h^3 in phase space. Any particle found in this volume will be assigned the coordinates of the phase point. The uncertainty in position and momentum implied by this procedure is therefore consistent with the uncertainty principle.

The cell functions, $\varphi(\mathbf{X}, \mathbf{K}, \mathbf{x})$, by virtue of the properties of the step function, provide an analytical means of dividing the configuration space into cells. They can be used to obtain an operator representation of the neutron field in the coarse-grained coordinates. Since these functions form an orthonormal and complete set, i.e.

$$\int d^3x\, \varphi^*(\mathbf{X}, \mathbf{K}, \mathbf{x})\, \varphi(\mathbf{X}', \mathbf{K}', \mathbf{x}) = \delta_{\mathbf{XX}'}\, \delta_{\mathbf{KK}'} \tag{2.6}$$

$$\sum_{\mathbf{X},\mathbf{K}} \varphi^*(\mathbf{X}, \mathbf{K}, \mathbf{x})\, \varphi(\mathbf{X}, \mathbf{K}, \mathbf{x}') = \delta(\mathbf{x} - \mathbf{x}') \tag{2.7}$$

the spinor field can be expanded as

$$\psi_j(\mathbf{x}) = \sum_{\mathbf{X},\mathbf{K},s} a(\mathbf{X}, \mathbf{K}, s)\, u_j(s)\, \varphi(\mathbf{X}, \mathbf{K}, \mathbf{x}) \tag{2.8}$$

where the functions $u_j(s)$, $s = \pm 1$, are the components of unit vectors in spin space which may have the simple representations,[2]

$$u(1) = \begin{pmatrix} 1 \\ 0 \end{pmatrix} \qquad u(-1) = \begin{pmatrix} 0 \\ 1 \end{pmatrix} \tag{2.9}$$

Furthermore, they have the properties that*

$$\sum_{s=\pm 1} u_j^+(s)\, u_k(s) = \delta_{jk} \tag{2.10}$$

$$u_j^+(s)\, u_j(s') = \delta_{ss'} \tag{2.11}$$

where the superscript "+" denotes Hermitian conjugate. Note that the index s labels the orientation of the neutron spin.

The coefficient in the expansion of $\psi_j(\mathbf{x})$,

$$a(\mathbf{X}, \mathbf{K}, s) = \int d^3x\, \varphi^*(\mathbf{X}, \mathbf{K}, \mathbf{x})\, u_j^+(s)\, \psi_j(\mathbf{x}) \tag{2.12}$$

is an operator governed by the same commutation relations specified for the field operator. For neutrons and other fermions the operators satisfy anticommutation rules,

$$[\psi_j(\mathbf{x}), \psi_k^+(\mathbf{x}')]_+ = \delta_{jk}\, \delta(\mathbf{x} - \mathbf{x}') \tag{2.13}$$

$$[\psi_j(\mathbf{x}), \psi_k(\mathbf{x}')]_+ = [\psi_j^+(\mathbf{x}), \psi_k^+(\mathbf{x}')]_+ = 0$$

* We employ the convention in which repeated spinor indices are summed.

Using Eq. 2.12 we find

$$[a(\mathbf{X}, \mathbf{K}, s), a^+(\mathbf{X}', \mathbf{K}', s')]_+ = \delta_{\mathbf{XX}'}\,\delta_{\mathbf{KK}'}\,\delta_{ss'}$$

$$[a(\mathbf{X}, \mathbf{K}, s), a(\mathbf{X}', \mathbf{K}', s')]_+ = [a^+(\mathbf{X}, \mathbf{K}, s), a^+(\mathbf{X}', \mathbf{K}', s')]_+ = 0 \quad (2.14)$$

For bosons Eq. 2.13 and 2.14 would still apply if everywhere the anti-commutator $[A, B]_+$ is replaced by the commutator $[A, B]$.

The operator $a(\mathbf{X}, \mathbf{K}, s)$ and its Hermitian conjugate are the conventional fermion destruction and creation operators. This is best illustrated by considering the effect of these operators when acting on a given state. Explicitly, let us consider an operator whose eigenvalue gives the total number of neutrons in a given state. This operator[2]

$$\mathcal{N} = \int d^3x\, \psi_j^+(\mathbf{x})\, \psi_j(\mathbf{x}) = \sum_{\mathbf{X}, \mathbf{K}, s} a^+(\mathbf{X}, \mathbf{K}, s)\, a(\mathbf{X}, \mathbf{K}, s) \quad (2.15)$$

is the total number operator and is term-wise Hermitian. A representation can always be found in which the operator

$$\varrho_1(\mathbf{X}, \mathbf{K}, s) = a^+(\mathbf{X}, \mathbf{K}, s)\, a(\mathbf{X}, \mathbf{K}, s) \quad (2.16)$$

is diagonal (hence $\mathcal{N}$ is diagonal since the ϱ's at different points commute), i.e.*

$$\varrho_1(\mathbf{X}, \mathbf{K}, s)\, |n\rangle = N(\mathbf{X}, \mathbf{K}, s)\, |n\rangle \quad (2.17)$$

where $N(\mathbf{X}, \mathbf{K}, s)$ is the number of neutrons at phase point $(\mathbf{X}, \hbar\mathbf{K})$ with spin orientation s. Since the neutron is a fermion, the occupation number $N(\mathbf{X}, \mathbf{K}, s)$ can be only zero or unity. The operator $\varrho_1(\mathbf{X}, \mathbf{K}, s)$ is seen to be the number operator at the indicated phase point.

In the above representation, the state $|n\rangle$ (ignoring other kinds of particles in the system for the moment) specifies the distribution of neutrons in **X**-**K**-s space as well as the total number of neutrons in the state. Thus

$$|n\rangle = |N(\mathbf{X}_1, \mathbf{K}_1, s_1)\, N(\mathbf{X}_1, \mathbf{K}_1, s_2) \ldots N(\mathbf{X}, \mathbf{K}, s) \ldots\rangle \quad (2.18)$$

with

$$\sum_{\mathbf{X}, \mathbf{K}, s} N(\mathbf{X}, \mathbf{K}, s) = N \quad (2.19)$$

where N is the corresponding eigenvalue of the total number operator $\mathcal{N}$. It will be convenient to replace the ordered arguments **X**, **K**, and s

* We use Dirac's notation of bras and kets.[3]

by an ordered set of subscripts with one-to-one correspondence. Eq. 2.18 then becomes

$$|n\rangle = |N_1 N_2 \dots N_\lambda \dots \rangle \tag{2.18a}$$

By using the appropriate commutation relations and starting with Eq. 2.17 one can readily show that for fermions[2]

$$a_\lambda |n\rangle = \theta_\lambda N_\lambda |N_1 N_2 \dots 1 - N_\lambda \dots\rangle \tag{2.20}$$

$$a_\lambda^+ |n\rangle = \theta_\lambda (1 - N_\lambda) |N_1 N_2 \dots N_\lambda \dots \rangle \tag{2.21}$$

$$\theta_\lambda = (-1)^{b_\lambda} \qquad b_\lambda = \sum_{l=1}^{\lambda-1} N_l$$

The phase factor θ_λ arises because the states before and after the operation of a_λ and a_λ^+ must be properly labeled. For bosons one finds

$$a_\lambda |n\rangle = [N_\lambda]^{1/2} |N_1 N_2 \dots N_\lambda - 1 \dots\rangle \tag{2.22}$$

$$a_\lambda^+ |n\rangle = [N_\lambda + 1]^{1/2} |N_1 N_2 \dots N_\lambda + 1 \dots\rangle \tag{2.23}$$

Of course, the occupation numbers for bosons can be any positive integer or zero.

B. A Kinetic Equation for $F(\mathbf{X}, \mathbf{K}, t)$

Having introduced the neutron number operator in coarse-grained phase space, we can now define a particle density, which has the same interpretation as that purportedly ascribed to $f(\mathbf{x}, \mathbf{v}, t)$ and which will be suitable for use in deriving an approximate transport equation for neutrons. Let the state of the system of interest at time t be denoted by $\Psi(t)$. The expected number of neutrons per unit cell volume at the phase point $(\mathbf{X}, \hbar\mathbf{K})$ is therefore given by

$$F(\mathbf{X}, \mathbf{K}, t) = L^{-3} \langle \Psi(t)| \varrho_1(\mathbf{X}, \mathbf{K}) |\Psi(t)\rangle \tag{2.24}$$

with

$$\varrho_1(\mathbf{X}, \mathbf{K}) = \sum_s a^+(\mathbf{X}, \mathbf{K}, s)\, a(\mathbf{X}, \mathbf{K}, s) \tag{2.25}$$

The expansion

$$\Psi(t) = \sum_n C_n(t) |n\rangle \tag{2.26}$$

results in another form of the expectation value

$$F(\mathbf{X}, \mathbf{K}, t) = L^{-3} \sum_{nm} D_{mn}(t) \langle n | \varrho_1(\mathbf{X}, \mathbf{K}) | m \rangle$$
$$= L^{-3} \operatorname{Tr} D(t) \varrho_1(\mathbf{X}, \mathbf{K}) \quad (2.27)$$

where

$$D_{mn}(t) = C_n^*(t) C_m(t) \quad (2.28)$$

is the von Neumann density matrix,[4,5] which is the quantum-mechanical equivalent of the classical Gibbs ensemble.*[5,6]

The time dependence of $F(\mathbf{X}, \mathbf{K}, t)$ is expressed through the density matrix operator which satisfies the quantum Liouville equation,[5]

$$\frac{\partial D}{\partial t} = \frac{i}{\hbar} [D, H] \quad (2.29)$$

H being the Hamiltonian of the system. It is worth noting that the trace is invariant under unitary transformations; hence, the representation in which Eq. 2.27 is evaluated may be chosen for convenience. Unless specifically stated otherwise, we shall calculate all matrix elements in the representation which diagonalizes the number operator. In the sense of Eq. 2.18, $D_{nn}(t)$ is seen to have the interpretation as the probability that at time t the system is in the state $|n\rangle$ in which the number of neutrons and their distributions in **X**-**K**-s space are specified.

The function $L^3 F(\mathbf{X}, \mathbf{K}, t)$ represents the expected number of neutrons with momentum $\mathbf{P} = \hbar \mathbf{K}$ and any spin orientation in the cell centered at **X** at time t. Since F is the expectation value of an operator whose eigenvalues are positive or zero, it is greater than or equal to zero everywhere and hence is appropriate as a particle distribution function. As defined, F is a density only in configuration space and not in momentum space; moreover, unlike the function f, it is not a distribution in continuous configuration space. The present derivation of the transport equation actually requires this discrete domain; however, since conventional results are usually expressed in a continuous momentum space, we will ultimately, whenever warranted, sum F over a small elemental volume d^3K according to

$$\sum_{\mathbf{K} \in d^3K} F(\mathbf{X}, \mathbf{K}, t) = \left(\frac{L}{2\pi}\right)^3 F(\mathbf{X}, \mathbf{K}, t) \, d^3K$$
$$= f(\mathbf{X}, \mathbf{K}, t) \, d^3K = f(\mathbf{X}, \mathbf{P}, t) \, d^3P \quad (2.30)$$

---

* The interpretation that a pure quantum-mechanical state corresponds to a classical ensemble is in agreement with van Kampen;[6a] see also Fano.[6b]

It is $f(\mathbf{X}, \mathbf{P}, t)$ that is to be identified as the analogue of the conventional neutron density.

It is perhaps of some value to digress and indicate briefly how the present approach is related to the phase-space distribution function employed in some recent investigations of transport phenomena.[1,7,8] Consider a generalized phase-space distribution function

$$g(\mathbf{x}, \mathbf{k}, t) = \int d^3y\, \varphi_k^*\left(\mathbf{x} - \frac{\mathbf{y}}{2}\right) \varrho^{(1)}\left(\mathbf{x} - \frac{\mathbf{y}}{2}, \mathbf{x} + \frac{\mathbf{y}}{2}, t\right) \varphi_k\left(\mathbf{x} + \frac{\mathbf{y}}{2}\right) \tag{2.31}$$

where $\{\varphi_k(\mathbf{x})\}$ is an orthonormal and complete set of space functions and $\varrho^{(1)}$ is a reduced density matrix given by

$$\varrho^{(1)}(\mathbf{x}, \mathbf{x}', t) = \mathrm{Tr}\, \psi_j^+(\mathbf{x}')\, \psi_j(\mathbf{x})\, D(t) \tag{2.32}$$

The function $g(\mathbf{x}, \mathbf{k}, t)$ has been studied by Mori[7] in deriving the Bloch equation,[9] and by Ono,[1] in the coarse-grained formalism, in deriving the Uehling-Uhlenbeck equation.[10] It provides a convenient means with which one can obtain either the fine-grained or the coarse-grained distribution functions. For if one uses plane wave for $\varphi_k(\mathbf{x})$, the result is equivalent to the familiar Wigner distribution function,[8,11]

$$g(\mathbf{x}, \mathbf{k}, t) = \int d^3y\, e^{-i\mathbf{k}\cdot\mathbf{y}}\, \varrho^{(1)}\left(\mathbf{x} - \frac{\mathbf{y}}{2}, \mathbf{x} + \frac{\mathbf{y}}{2}, t\right) \tag{2.33}$$

If the cell function is used the result is

$$g(\mathbf{x}, \mathbf{X}, \mathbf{K}, t) = \int d^3y\, \varphi^*\left(\mathbf{X}, \mathbf{K}, \mathbf{x} - \frac{\mathbf{y}}{2}\right) \varrho^{(1)}\left(\mathbf{x} - \frac{\mathbf{y}}{2}, \mathbf{x} + \frac{\mathbf{y}}{2}, t\right) \times \\ \times \varphi\left(\mathbf{X}, \mathbf{K}, \mathbf{x} + \frac{\mathbf{y}}{2}\right) \tag{2.34}$$

The coarse-grained distribution function is then obtained by integrating $g(\mathbf{x}, \mathbf{X}, \mathbf{K}, t)$,

$$\begin{aligned} G(\mathbf{X}, \mathbf{K}, t) &= \int d^3x\, g(\mathbf{x}, \mathbf{X}, \mathbf{K}, t) \\ &= \int d^3x\, d^3x'\, \varphi^*(\mathbf{X}, \mathbf{K}, \mathbf{x})\, \varphi(\mathbf{X}, \mathbf{K}, \mathbf{x}')\, \mathrm{Tr}\, \psi_j^+(\mathbf{x})\, \psi_j(\mathbf{x}')\, D(t) \end{aligned} \tag{2.35}$$

In view of the spinor field expansion, Eq. 2.8, the above expression for $G(\mathbf{X}, \mathbf{K}, t)$ is seen to differ from Eq. 2.27 only by a volume factor.

We now consider the time dependence of $F(\mathbf{X}, \mathbf{K}, t)$. If the system Hamiltonian is assumed not to be an explicit function of time,* then a formal solution to the operator equation, Eq. 2.29, is

$$D(t + \tau) = e^{-i\tau H/\hbar} D(t) e^{i\tau H/\hbar} \tag{2.36}$$

The Hamiltonian can be written quite generally as

$$H = T + H_s + V + V^{nn} \tag{2.37}$$

where T is the kinetic energy of the neutrons, H_s is that part of the energy of the system independent of the presence or absence of neutrons,† V is the energy of neutron-nuclear and photon-nuclear interactions and V^{nn} is the energy of the neutron-neutron interaction. In the following we shall ignore V^{nn}, as its effects are truly negligible in the studies of neutron transport in macroscopic media.‡

In the quantized field formalism the nonrelativistic neutron kinetic energy is of the form[2]

$$T = -\frac{\hbar^2}{2m} \int d^3x \, \psi_j^+(\mathbf{x}) \nabla^2 \psi_j(\mathbf{x}) \tag{2.38}$$

where m is the neutron mass. This operator can be expressed in terms of coarse-grained coordinates by means of the spinor field expansion. We obtain

$$T = \mathscr{E} + T' + T'' \tag{2.39}$$

---

* Problems with time-dependent Hamiltonians are also of general interest.[12] In the present case it might be realized if the neutrons were exposed to a time-varying gravitational or inhomogeneous magnetic field. These effects, however, are not likely to be significant in neutron transport theory.

† In passing, we observe that if H_s is made sufficiently inclusive and if an appropriate selection of operator representatives of dynamical variables to be measured is made in any given case, then Eq. 2.29 along with the general form of an observable expectation value, $\omega(t) = \mathrm{Tr}\, \Omega\, D(t)$, encompass Maxwell's equations—hence all of classical electricity and magnetism; equations for radiant energy (photon) transfer[13,15]—hence all of the equations of reactor shielding as well as the theory of photon interactions with matter; equations for neutral[16] and charged particle[12] gas kinetics; equations of Newton—hence all of classical mechanics; etc. This is merely an involved way of suggesting that, in our opinion, Eq. 2.29 and $\omega(t)$, when appropriately phrased, provide a suitable starting point for investigations in virtually all branches of science and engineering.

‡ The ratio of the neutron density to the nuclear density in a reactor is at most 10^{-7} or less. If one assumes that the cross section for (n, n) scattering is roughly the same as that for (n, p), then the mean free path for neutron-neutron interaction is of order 10^8 cm or more.

where

$$\mathscr{E} = \sum_{\mathbf{X},\mathbf{K},s} \frac{\hbar^2 K^2}{2m} \varrho_1(\mathbf{X}, \mathbf{K}, s) \tag{2.40a}$$

$$T' = \frac{i\hbar^2}{2mL^3} \sum_{\substack{\mathbf{X}',\mathbf{K}',s\\ \mathbf{X},\mathbf{K}}} a^+(\mathbf{X}, \mathbf{K}, s)\, a(\mathbf{X}', \mathbf{K}', s) \int d^3x\, e^{i\mathbf{x}\cdot(\mathbf{K}'-\mathbf{K})} \times$$

$$\times\, [E(\mathbf{X}', \mathbf{x})\, \mathbf{K}' \cdot \nabla E(\mathbf{X}, \mathbf{x}) - E(\mathbf{X}, \mathbf{x})\, \mathbf{K} \cdot \nabla E(\mathbf{X}', \mathbf{x})] \tag{2.40b}$$

$$T'' = \frac{\hbar^2}{2mL^3} \sum_{\substack{\mathbf{X}',\mathbf{K}',s\\ \mathbf{X},\mathbf{K}}} a^+(\mathbf{X}, \mathbf{K}, s)\, a(\mathbf{X}', \mathbf{K}', s) \int d^3x\, e^{i\mathbf{x}\cdot(\mathbf{K}'-\mathbf{K})} \times$$

$$\times\, [\nabla E(\mathbf{X}, \mathbf{x}) \cdot \nabla E(\mathbf{X}', \mathbf{x})] \tag{2.40c}$$

The term $\mathscr{E}$ represents the sum of neutron kinetic energies at every phase point. This term will serve the useful purpose of determining the neutron states between which collision-induced transitions take place. The term T', as will be seen presently, describes the transport of neutrons from cell to cell. The term T'' represents an apparent infinite contribution to the Hamiltonian. It is surmised that this term is actually meaningless and should henceforth be ignored.*

The Hamiltonian now appears as

$$H = T' + H' \tag{2.41}$$

with

$$H' = \mathscr{E} + H_s + V \tag{2.42}$$

This particular decomposition of H is made because we anticipate that H' is important in connection with collision processes only. In general it is not true that V is concerned solely with the effect of collisions on the variation of F. If the particles interact with "external" force fields, or with each other or other kinds of particles through forces characterized by effective ranges substantially greater than L, then a portion of V should be incorporated into the description of "transport".[12] The long-range part of V will then provide smoothly varying forces leading

* Keeping T'' would mean considering the expectation value Tr $D[T'',\varrho]$ (cf. Eq. 2.50a). To the same order of approximation made in the subsequent analysis of the effects of particle streaming (Section C) one can demonstrate that the commutator, $[T'', a^+(\mathbf{X}, \mathbf{K})\, a(\mathbf{X}, \mathbf{K})]$, vanishes identically. Thus, for the purpose of the present derivation the term T'' actually gives no contribution in the description of transport.

to curved trajectories for the particles between impulsive events. For the present discussion, however, we shall assume that V represents only extremely short-range interactions. The operators T' and H' therefore will give rise to transport and collision phenomena respectively.

It will be desirable to treat the effects of transport and collisions separately. To do this we first write

$$U(\tau) = e^{-i\tau(T'+H')/\hbar} = e^{-i\tau T'/\hbar}\, e^{-i\tau H'/\hbar}\, J(\tau) \tag{2.43}$$

where $J(\tau)$ is a unitary operator to be determined by the equation,

$$\frac{\partial J}{\partial t} = u(t)\, J \tag{2.44}$$

and the boundary condition $J(0) = I$, I being the identity operator. The function $u(t)$ can be represented in a variety of ways. Two such examples are

$$u(t) = \frac{i}{\hbar}\,[H' - U'^{+}(t)\, e^{itT'/\hbar}\, H'\, e^{-itT'/\hbar}\, U'(t)] \tag{2.45}$$

and

$$u(t) = -\frac{i}{\hbar} \sum_{\substack{n,m=0\\ n+m\neq 0}}^{\infty} \frac{(it/\hbar)^{n+m}}{n!\, m!}\, [H', [T', H']_n]_m \tag{2.46}$$

where

$$U'(t) = e^{-itH'/\hbar} \tag{2.47}$$

and $[A, B]_n$ denotes the nth order commutator of A and B, i.e.

$$\begin{aligned} [A, B]_0 &= B \\ [A, B]_1 &= [A, B] \\ [A, B]_2 &= [A, [A, B]] \end{aligned} \tag{2.48}$$

etc. Making use of this operator decomposition in Eq. 2.36, we find that Eq. 2.27 becomes

$$\begin{aligned} F(t+\tau) &= L^{-3}\, \mathrm{Tr}\, \varrho_1 U(\tau)\, D(t)\, U^{+}(\tau) \\ &= L^{-3}\, \mathrm{Tr}\, e^{i\tau T'/\hbar}\, \varrho_1\, e^{-i\tau T'/\hbar}\, U'(\tau)\, J(\tau)\, D(t)\, J^{+}(\tau)\, U'^{+}(\tau) \end{aligned} \tag{2.49}$$

where use is made of the cyclic invariance of the trace. The dependence of F and ϱ_1 on $\mathbf{X}$ and $\mathbf{K}$ will not be explicitly indicated when no risk of confusion is incurred.

Thus far we have proceeded formally without considering the structure of the equation we ultimately wish to obtain. The fact that the neutron transport equation, Eq. 1.1, is a first-order linear differential equation in time suggests that, to derive a similar equation for F, Eq. 2.49 should be examined for some small time interval, τ. One can anticipate that there will be a range of intervals, say $\tau_1 < \tau < \tau_2$, in which it is meaningful to decompose Eq. 2.49 into terms describing either transport or collision effects. The description of transport is expected to be valid so long as τ is less than some upper limit τ_2, whereas the description of collisions is expected to be valid for τ greater than some lower limit τ_1. These limits are rather ill-defined at this point, but a qualitative estimate for τ_2 is suggested by the requirement that $[(\partial^2 F/\partial t^2)/(\partial F/\partial t)] \ll \tau_2^{-1}$, and for τ_1 one may take the neutron-nuclear interaction times to be discussed later.

According to the above considerations we will treat τ as a small but finite interval. Then

$$F(t+\tau) = L^{-3}\,\mathrm{Tr}\left\{\varrho_1 + \frac{i\tau}{\hbar}[T', \varrho_1] + \sum_{m=2}^{\infty} \frac{(i\tau/\hbar)^m}{m!}[T', \varrho_1]_m\right\} \times$$

$$\times\; U'(\tau)\, J(\tau)\, D(t)\, J^{+}(\tau)\, U'^{+}(\tau)$$

$$\approx L^{-3}\,\mathrm{Tr}\left\{\varrho_1 + \frac{i\tau}{\hbar}[T', \varrho_1]\right\} U'(\tau)\, J(\tau)\, D(t)\, J^{+}(\tau)\, U'^{+}(\tau) \qquad (2.50)$$

For sufficiently small τ such that all terms in the m sum can be neglected, transport is described by the second term in Eq. 2.50. Since this term is already proportional to τ we will keep only the leading term in the transformation

$$U'(\tau)\, J(\tau)\, D(t)\, J^{+}(\tau)\, U'^{+}(\tau) = D(t) + O(\tau) \qquad (2.51)$$

For the first term in Eq. 2.50 we will ignore the effect of the operator J. This approximation is justified by the fact that

$$J(\tau) = I + O(\tau^2) \qquad (2.52)$$

In more physical terms the neglect of J here implies that in treating collisions in a given cell the effects due to particles outside the cell are ignored. Eq. 2.50 now becomes

$$F(t+\tau) \approx L^{-3}\,\mathrm{Tr}\left\{\varrho_1\, U'(\tau)\, D(t)\, U'^{+}(\tau) + \frac{i\tau}{\hbar}[T', \varrho_1]\, D(t)\right\} \qquad (2.50a)$$

The present approximations result in a complete separation of the effects of T' and H', and hence will lead to a transport equation in which terms affected by transport or collision processes enter independently of each other. This does not mean, however, that the momentum and spatial dependence of the solution is decoupled.

In the representation which diagonalizes ϱ_1 the first term in Eq. 2.50a may be arranged to give

$$\operatorname{Tr} \varrho_1\, U'(\tau)\, D(t)\, U'^{+}(\tau) \approx \sum_{nn'} N(\mathbf{X}, \mathbf{K})\, D_{nn}(t)\, |U'_{nn'}(\tau)|^2 \tag{2.53}$$

where we have ignored the off-diagonal elements of the density matrix.* Since $U'(\tau)$ is unitary,

$$|U'_{nn}(\tau)|^2 = 1 - \sum_{n'}{}' \, |U'_{n'n}(\tau)|^2 \tag{2.54}$$

where the prime on the summation indicates that terms for which $n = n'$ are to be excluded. Then Eq. 2.53 becomes

$$\operatorname{Tr} \varrho_1\, U'(\tau)\, D(t)\, U'^{+}(\tau)$$

$$\approx L^3 F(t) + \sum_{nn'} D_{nn}(t)\, |U'_{n'n}(\tau)|^2\, [N'(\mathbf{X}, K) - N(\mathbf{X}, \mathbf{K})] \tag{2.55}$$

Here the occupation numbers N' and N denote the eigenvalues of ϱ_1 in the states $|n'\rangle$ and $|n\rangle$ respectively. Combining Eq. 2.50a and 2.55 we have

$$[F(\mathbf{X}, \mathbf{K}, t + \tau) - F(\mathbf{X}, \mathbf{K}, t)]\, \tau^{-1} - \frac{i}{\hbar L^3} \operatorname{Tr}\, [T', \varrho_1(\mathbf{X}, \mathbf{K})]\, D(t)$$

$$\approx L^{-3} \sum_{nn'} W_{n'n}(\tau)\, D_{nn}(t)\, [N'(\mathbf{X}, \mathbf{K}) - N(\mathbf{X}, \mathbf{K})] \tag{2.56}$$

where $\qquad W_{n'n}(\tau) = |U'_{n'n}(\tau)|^2/\tau$

The first term will be replaced by the time derivative of F, although in a strict sense it should always be thought of as a finite difference. However, even for $\tau_1 \approx 10^{-8}$ sec., coarse-graining of the time domain is not likely to be significant in most investigations of physical systems.

The remainder of this chapter will be devoted to a reduction of the remaining terms in Eq. 2.56. It will be shown that the second term provides the conventional description of neutron transport, whereas the terms on the right-hand side provide the description of interactions. With these reductions, Eq. 2.56 will then bear considerable similarity to Eq. 1.1.

---

* This is equivalent to the so-called Random *a priori* Phase Approximation which has been studied only in very special cases.[17]

C. The Streaming Term

In order to exhibit in Eq. 2.56 the transport term that appears in Eq. 1.1 it is necessary to evaluate the commutator,

$$[T', \varrho_1(\mathbf{X}, \mathbf{K})] = \frac{i\hbar^2}{2mL^3} \sum_{\substack{\mathbf{X}', \mathbf{X}'', \mathbf{K}' \\ \mathbf{K}'', s'}} [a^+(\mathbf{X}', \mathbf{K}', s')\, a(\mathbf{X}'', \mathbf{K}'', s'), \varrho_1(\mathbf{X}, \mathbf{K})] \times$$
$$\times \int \mathrm{d}^3x\, \mathrm{e}^{i\mathbf{x}\cdot(\mathbf{K}''-\mathbf{K}')} [E(\mathbf{X}'', \mathbf{x})\, \mathbf{K}'' \cdot \nabla E(\mathbf{X}', \mathbf{x})$$
$$- E(\mathbf{X}', \mathbf{x})\, K \cdot \nabla E(\mathbf{X}'', x)] \tag{2.57}$$

One can readily show, using the commutation relations given in Eq. 2.14, that

$$[a^+(\mathbf{X}', \mathbf{K}', s')\, a(\mathbf{X}'', \mathbf{K}'', s'), a^+(\mathbf{X}, \mathbf{K}, s)\, a(\mathbf{X}, \mathbf{K}, s)]$$
$$= \{a^+(\mathbf{X}', \mathbf{K}', s)\, a(\mathbf{X}, \mathbf{K}, s)\, \delta_{\mathbf{X}\mathbf{X}''}\, \delta_{\mathbf{K}\mathbf{K}''}$$
$$- a^+(\mathbf{X}, \mathbf{K}, s)\, a(\mathbf{X}'', \mathbf{K}'', s)\, \delta_{\mathbf{X}\mathbf{X}'}\, \delta_{\mathbf{K}\mathbf{K}'}\}\, \delta_{ss'} \tag{2.58}$$

Thus Eq. 2.57 assumes the form,

$$[T', \varrho_1(\mathbf{X}, \mathbf{K})] = \Omega - \Omega^+ \tag{2.59}$$

where

$$\Omega = \frac{i\hbar^2}{2mL^3} \sum_{\mathbf{X}', \mathbf{K}', s} a^+(\mathbf{X}', \mathbf{K}', s)\, a(\mathbf{X}, \mathbf{K}, s) \int \mathrm{d}^3x\, \mathrm{e}^{i\mathbf{x}\cdot(\mathbf{K}-\mathbf{K}')} \times$$
$$\times [E(\mathbf{X}, \mathbf{x})\, \mathbf{K} \cdot \nabla E(\mathbf{X}', \mathbf{x}) - E(\mathbf{X}', \mathbf{x})\, \mathbf{K}' \cdot \nabla E(\mathbf{X}, \mathbf{x})] \tag{2.60}$$

With the help of Eqs. 2.4 and 2.5 we find

$$\Omega = \frac{i\hbar^2}{mL^3} \sum_{\mathbf{X}', \mathbf{K}', s} a^+(\mathbf{X}', \mathbf{K}', s)\, a(\mathbf{X}, \mathbf{K}, s) \int \mathrm{d}^3x\, E(\mathbf{X}, \mathbf{x})\, E(\mathbf{X}', \mathbf{x})\, \mathrm{e}^{i\mathbf{x}\cdot(\mathbf{K}-\mathbf{K}')} \times$$
$$\times \sum_{j=1}^{3} \left\{ K_j \left[\delta\left(x_j - X_j' + \frac{L}{2}\right) - \delta\left(x_j - X_j' - \frac{L}{2}\right) \right] \right.$$
$$\left. - K_j' \left[\delta\left(x_j - X_j + \frac{L}{2}\right) - \delta\left(x_j - X_j - \frac{L}{2}\right) \right] \right\}$$
$$= \frac{i\hbar^2}{4mL^3} \sum_{s,j} \sum_{K_j'} (\pm)\, (K_j + K_j') \times$$
$$\times \{[a^+(\mathbf{X}_j + L, K_j') - a^+(X_j - L, K_j')]\, a(X_j, K_j)\} \tag{2.61}$$

where the upper or lower sign is used depending upon whether $M_j = (L/2\pi)(K_j - K_j')$ is even or odd respectively in the K_j' sum. In writing the arguments of the operators we have suppressed the spin labels as well as those components of the two phase points $(\mathbf{X}', \hbar\mathbf{K}')$ and $(\mathbf{X}, \hbar\mathbf{K})$ which are the same.

The contributions from the $\mathbf{K} = \mathbf{K}'$ terms in Ω and Ω^+ are easily recognized. Taking only these terms into account the transport term in Eq. 2.56 becomes

$$-\frac{i}{\hbar L^3}\,\mathrm{Tr}\,[T', \varrho_1(\mathbf{X}, \mathbf{K})]\,D(t) = \frac{\hbar}{m}\,\mathbf{K}\cdot\nabla F(\mathbf{X}, \mathbf{K}, t) \qquad (2.62)$$

The gradient operation actually represents a finite difference in the sense that

$$\frac{\partial F(\mathbf{X}, \mathbf{K}, t)}{\partial X_1} = L^{-3} \times$$

$$\times\,\mathrm{Tr}\sum_s \{[a^+(X_1 + L, X_2, X_3, \mathbf{K}, s) - a^+(X_1 - L, X_2, X_3, \mathbf{K}, s)]\,a(\mathbf{X}, \mathbf{K}, s)$$

$$+\,a^+(\mathbf{X}, \mathbf{K}, s)\,[a(X_1 + L, X_2, X_3, \mathbf{K}, s) - a(X_1 - L, X_2, X_3, \mathbf{K}, s)]\}\times$$

$$\times\,D(t)/2L \qquad (2.63)$$

Eq. 2.62 is seen to be the conventional streaming term which rigorously describes the flow of nonrelativistic, massive particles in free space in both classical and fine-grained quantum theories.

The contribution from $K \neq K'$ terms in Ω do not lead to any readily interpretable result. However, they appear to describe the correlation of neutrons with different momenta in adjacent cells. Such effects may be regarded as corrections to the streaming term due to space-momentum coupling. Because these terms do not appear in a fine-grained theory, it is reasonable to conclude that the coupling is a direct consequence of coarse-graining. Indeed, as L becomes arbitrarily small, the separations between $\hbar\mathbf{K}$, the momentum point under consideration, and other points in the momentum lattice approach infinity. One may then anticipate the $\mathbf{K}'$ sum to collapse to only the $\mathbf{K} = \mathbf{K}'$ term.

It would be of interest to investigate the quantitative effects of these terms. We observe that a typical term in Ω is

$$\mathrm{Tr}\,a^+(X_j + L, K_j')\,a(X_j, K_j)\,D(t)$$

$$= \langle\Psi(t)|\,a^+(X_j + L, K_j')\,a(X_j, K_j)\,|\Psi(t)\rangle \qquad (2.64)$$

which has the appearance of a reduced density in coarse-grained phase-space coordinates (see Eq. 2.32). Similar quantities have been encountered in recent studies of many-body problems such as ground state energies, the nature of elementary excitations, and thermodynamics.[18] Thus the formalism and the techniques developed for those approaches[18] to a statistical theory of interacting particles may well be applied in the present context to the understanding of the $K \neq \mathbf{K}'$ terms.

D. The Collision Terms

We have derived in Eq. 2.56 a kinetic equation for the coarse-grained neutron density $F(\mathbf{X}, \mathbf{K}, t)$. The effects of neutron-nuclear interactions are described by the transition probability per unit time, $W_{n'n}$. With regard to the neutron transport equation the reactions of primary interest are capture, scattering, and fission. A direct calculation of the transition probabilities, hence the cross sections, associated with these processes is a rather involved task and will be considered in the next two chapters. Our main concern here, therefore, is to extract the dependence on neutron density of the various relevant collision contributions so that Eq. 2.56 can be directly compared to Eq. 1.1.

It is convenient to subdivide the elements of the transition matrix W into classes according to the number of neutrons in the state $|n'\rangle$ as compared to the number in the state $|n\rangle$. The reason is that given a fixed number in the initial state the number in the final state depends upon whether the reaction is capture, scattering, or fission, and is different in all three cases. A possible ambiguity may arise in the case of a fission event producing only one neutron. This particular case, although indistinguishable in the present context from a similar situation in scattering, is actually different when the nuclei involved are taken into consideration. Thus the sum over n', which may be regarded as the final state, can be exhibited as a sum of distinct contributions corresponding to the three types of interactions.*

Consider first the capture of a neutron with wave vector $\mathbf{K}_j$ in the space cell $\mathbf{X}_i$. The transition probability for this reaction is simply

$$W^c_{n'n} = \overline{w}^c_{\mathbf{K}_j} N(\mathbf{X}_i, \mathbf{K}_j) \tag{2.65}$$

* All interactions are treated as binary collisions.

where $\overline{w}^c_{\mathbf{K}_j}$ is the reduced transition probability for the capture of a neutron at phase point $(\mathbf{X}_i, h\mathbf{K}_j)$.* Note that if initially there is no neutron at $(\mathbf{X}, \mathbf{K})$, the present interaction would have zero contribution. The sum over n' in this case implies a sum over all states in which the total neutron number is one less than the number in the state $|n\rangle$. It is effectively a sum over all $\mathbf{X}_i$ and $\mathbf{K}_j$; however, because of the factor $[N'(\mathbf{X}, \mathbf{K}) - N(\mathbf{X}, \mathbf{K})]$ all terms would vanish unless $\mathbf{X}_i = \mathbf{X}$ and $\mathbf{K}_j = \mathbf{K}$, in which case the factor becomes -1. Hence,

$$\begin{aligned} I_c &= L^{-3} \sum_{nn'} D_{nn} W_{n'n} [N'(\mathbf{X}, \mathbf{K}) - N(\mathbf{X}, \mathbf{K})] \\ &= -\overline{w}^c_{\mathbf{K}} F(\mathbf{X}, \mathbf{K}, t) \end{aligned} \tag{2.66}$$

Consider next the scattering† of a neutron in cell $\mathbf{X}_i$ from initial wave vector $\mathbf{K}_j$ to final wave vector $\mathbf{K}_l$. This process is equivalent to the absorption of a neutron at $(\mathbf{X}_i, \mathbf{K}_j)$ and the creation of one at $(\mathbf{X}_i, \mathbf{K}_l)$, so that

$$W^s_{n'n} = \overline{w}^s_{\mathbf{K}_j \to \mathbf{K}_l} N(\mathbf{X}_i, \mathbf{K}_j) \left[1 - \frac{N(\mathbf{X}_i, \mathbf{K}_l)}{2} \right] \tag{2.67}$$

where $\overline{w}^s_{\mathbf{K}_j \to \mathbf{K}_l}$ is the reduced transition probability for the scattering of a neutron in cell $\mathbf{X}_i$ from $\mathbf{K}_j$ to $\mathbf{K}_l$.‡ The factor $[1 - N(\mathbf{X}_i, \mathbf{K}_l)/2]$ is the number dependence associated with the creation process and we have assumed that the neutron spin orientation is random, i.e.

$$N(\mathbf{X}, \mathbf{K}, s) = N(\mathbf{X}, \mathbf{K})/2 \tag{2.68}$$

with $N(\mathbf{X}, \mathbf{K})$ equal to zero, one or two. The factor $[N'(\mathbf{X}, \mathbf{K}) - N(\mathbf{X}, \mathbf{K})]$ can be either $+1$ or -1 in this case depending upon whether $(\mathbf{X}_i, \mathbf{K}_j) = (\mathbf{X}, \mathbf{K})$ or $(\mathbf{X}_i, \mathbf{K}_l) = (\mathbf{X}, \mathbf{K})$. Thus we find

$$\begin{aligned} L^3 I_s &= \sum_{nn'} D_{nn} W^s_{n'n} [N'(\mathbf{X}, \mathbf{K}) - N(\mathbf{X}, \mathbf{K})] \\ &= -\sum_{n} D_{nn} N(\mathbf{X}, \mathbf{K}) \sum_{\mathbf{K}_l} \overline{w}^s_{\mathbf{K} \to \mathbf{K}_l} \left[1 - \frac{N(\mathbf{X}, \mathbf{K}_l)}{2} \right] \\ &\quad + \sum_{n} D_{nn} \left[1 - \frac{N(\mathbf{X}, \mathbf{K})}{2} \right] \sum_{\mathbf{K}_j} \overline{w}^s_{\mathbf{K}_j \to \mathbf{K}} N(\mathbf{X}, \mathbf{K}_j) \end{aligned} \tag{2.69}$$

---

* Spatial dependence of $\overline{w}$ will be suppressed for simplicity.

† The distinction between elastic and inelastic scattering is not necessary here. In the next chapter, the two processes will have to be treated separately.

‡ We ignore here any spin-dependent effects in the scattering.

It is obvious that the loss term in I_s represents the scattering of a neutron out of the phase point $(\mathbf{X}, \hbar\mathbf{K})$ while the gain term represents the scattering of a neutron into $(\mathbf{X}, \hbar\mathbf{K})$. Whenever $\mathbf{K}_j = \mathbf{K}_l = \mathbf{K}$ the net contribution vanishes as expected.

In Eq. 2.69 we have terms proportional to the expectation value of a product of two number operators which is a higher-order density and can be defined as

$$F_2(\mathbf{X}, \mathbf{K}, \mathbf{X}', \mathbf{K}', t) = L^{-6} \operatorname{Tr} \varrho_1(\mathbf{X}, \mathbf{K})\, \varrho_1(\mathbf{X}', \mathbf{K}',)\, D(t) \qquad (2.70)$$

It is conventional to call $F(\mathbf{X}, \mathbf{K}, t)$ a singlet density and $F_2(\mathbf{X}, \mathbf{K}, \mathbf{X}'\, \mathbf{K}', t)$ a doublet density. The appearance of the doublet is a consequence of the quantum statistics, and hence these terms can be expected to vanish in the classical limit. To show this we need to transform Eq. 2.69 to continuous momentum space according to Eq. 2.30. It is found that

$$\sum_{\mathbf{K}\in \mathrm{d}^3K} I_s = \mathscr{I}_s(\mathbf{K})\, \mathrm{d}^3K = \mathscr{I}_s(\mathbf{P})\, \mathrm{d}^3P \qquad (2.71)$$

and

$$\begin{aligned} \mathscr{I}_s(P) = &-f(\mathbf{X}, \mathbf{P}, t) \int \mathrm{d}^3P_l\, w_s(\mathbf{P} \to \mathbf{P}_l) \\ &+ \int \mathrm{d}^3P_j\, f(\mathbf{X}, \mathbf{P}_j, t)\, w_s(\mathbf{P}_j \to \mathbf{P}) \\ &+ \frac{(2\pi\hbar)^3}{2} \int \mathrm{d}^3P_l\, f_2(\mathbf{X}, \mathbf{P}, \mathbf{X}, \mathbf{P}_l, t)\, w_s(\mathbf{P} \to \mathbf{P}_l) \\ &- \frac{(2\pi\hbar)^3}{2} \int \mathrm{d}^3P_j\, f_2(\mathbf{X}, \mathbf{P}, \mathbf{X}, \mathbf{P}_j, t)\, w_s(\mathbf{P}_j \to \mathbf{P}) \end{aligned} \qquad (2.72)$$

Observe that $\overline{w}^s_{\mathbf{K}\to\mathbf{K}_l}$, when we summed over the elemental volume d^3K_l, becomes a distribution,

$$\sum_{\mathbf{K}_l\in \mathrm{d}^3K_l} \overline{w}^s_{\mathbf{K}\to\mathbf{K}_l} = w_s(\mathbf{K} \to \mathbf{K}_l)\, \mathrm{d}^3K_l = w_s(\mathbf{P} \to \mathbf{P}_l)\, \mathrm{d}^3P_l \qquad (2.73)$$

Thus in the classical limit ($\hbar \to 0$) only the first two terms in Eq. 2.72 survive.

Lastly we consider the contributions arising from fission processes. Let a neutron be absorbed at $(\mathbf{X}_i, \mathbf{K}_j)$ and the resulting reaction produce J neutrons with momentum distribution specified by a set of wave vectors, $\{K_l\}_J$. We shall neglect delayed effects so these neutrons are all emitted in the spatial cell $\mathbf{X}_i$ and within the time interval τ_1. The transition probability for this event can be written as

$$W^F_{n'n} = \overline{w}^F_{\mathbf{K}_j\to\{\mathbf{K}_l\}_J}\, N(\mathbf{X}_i, \mathbf{K}_j)\, G(\mathbf{X}, \{\mathbf{K}_l\}_J) \qquad (2.74)$$

where $\overline{w}^F_{\mathbf{K}_j \to \{\mathbf{K}_l\}_J}$ is the reduced fission transition probability, and

$$G(\mathbf{X}_i, \{K_l\}_J) = \prod_{\{\mathbf{K}_l\}_J} \left[1 - \frac{N(\mathbf{X}_i, \mathbf{K}_l)}{2} \right] \tag{2.75}$$

is the degeneracy factor which contains a product of J factors according to the J wave vectors in $\{\mathbf{K}_l\}_J$. It has just been shown that these terms lead to a dependence on higher-order densities which vanishes as $\hbar \to 0$. Since we are primarily interested in the fission contributions in the classical limit we shall replace G by unity in the following. Hence

$$\begin{aligned}
L^3 I_F = &- \sum_n D_{nn} N(\mathbf{X}, \mathbf{K}) \sum_{J, \{\mathbf{K}_l\}_J} \overline{w}^F_{\mathbf{K} \to \{\mathbf{K}_l\}_J} \left(1 - \sum_{\alpha=1}^{J} Q^\alpha_{\mathbf{K}\{\mathbf{K}_l\}_J} \right) \\
&+ \sum_n D_{nn} \sum_{\substack{J, \{\mathbf{K}_l\}_J \\ \mathbf{K}_j}} \overline{w}^F_{\mathbf{K}_j \to \{\mathbf{K}_l\}_J} N(\mathbf{X}, \mathbf{K}_j) (1 - \delta_{\mathbf{K}\mathbf{K}_j}) \sum_{\alpha=1}^{J} \alpha Q^\alpha_{\mathbf{K}\{\mathbf{K}_l\}_J} \\
&+ \sum_n D_{nn} N(\mathbf{X}, \mathbf{K}) \sum_{J, \{\mathbf{K}_l\}_J} \overline{w}^F_{\mathbf{K} \to \{\mathbf{K}_l\}_J} \sum_{\alpha=1}^{J} (\alpha - 1) Q^\alpha_{\mathbf{K}\{\mathbf{K}_l\}_J} \\
= &- \sum_n D_{nn} N(\mathbf{X}, \mathbf{K}) \sum_{J, \{\mathbf{K}_l\}_J} \overline{w}^F_{\mathbf{K}_j \to \{\mathbf{K}_l\}_j} \\
&+ \sum_n D_{nn} \sum_{\substack{J, \{\mathbf{K}_l\}_J \\ \mathbf{K}_j}} N(\mathbf{X}, \mathbf{K}_j) \overline{w}^F_{\mathbf{K}_j \to \{\mathbf{K}_l\}_j} \sum_{\alpha=1}^{J} \alpha Q^\alpha_{\mathbf{K}\{\mathbf{K}_l\}_J}
\end{aligned} \tag{2.76}$$

where $Q^\alpha_{\mathbf{K}\{\mathbf{K}_l\}_J}$ is the probability that of the J neutrons emitted with momentum distribution $\{\mathbf{K}_l\}_J$ there are exactly α neutrons with wave vector $\mathbf{K}$.

The collision terms, Eqs. 2.66, 2.69, and 2.76, along with the transport term, Eq. 2.62, can be entered into Eq. 2.56. Keeping in mind that time and spatial derivatives are actually finite differences, we may exhibit the resulting equation, in the absence of quantum effects, as

$$\begin{aligned}
&\left(\frac{\partial}{\partial t} + \frac{\hbar}{m} \mathbf{K} \cdot \nabla_X + \overline{w} \right) F(\mathbf{X}, \mathbf{K}, t) \\
&\quad = \sum_{\mathbf{K}_j} F(\mathbf{X}, \mathbf{K}_j, t) \left[\overline{w}^s_{\mathbf{K}_j \to \mathbf{K}} + \sum_{J, \{\mathbf{K}_l\}_J} \overline{w}^F_{\mathbf{K}_j \to \{\mathbf{K}_l\}_J} \sum_{\alpha=1}^{J} \alpha Q^\alpha_{\mathbf{K}\{\mathbf{K}_l\}_J} \right]
\end{aligned} \tag{2.77}$$

where

$$\overline{w} = \overline{w}^c_{\mathbf{K}} + \sum_{\mathbf{K}_l} \overline{w}^s_{\mathbf{K} \to \mathbf{K}_l} + \sum_{J, \{\mathbf{K}_l\}_J} \overline{w}^F_{\mathbf{K} \to \{\mathbf{K}_l\}_J} = \overline{w}^c_{\mathbf{K}} + \overline{w}^s_{\mathbf{K}} + \overline{w}^F_{\mathbf{K}} \tag{2.78}$$

For this equation to be directly comparable to the conventional neutron transport equation, it is necessary that we transform to continuous momentum space and express the transition probabilities as cross sections. The present transport equation then becomes

$$\left[\frac{\partial}{\partial t} + \frac{\mathbf{P}}{m} \cdot \nabla + \frac{P}{m} \Sigma_t (\mathbf{P})\right] f(\mathbf{X}, \mathbf{P}, t)$$

$$= \int \mathrm{d}^3P' \frac{P'}{m} f(\mathbf{X}, \mathbf{P}', t) \left[\Sigma_s(\mathbf{P}') \mathscr{F}(\mathbf{P}' \to \mathbf{P}) + \Sigma_f(\mathbf{P}') \sum_{\mathbf{J}, \alpha} \alpha B^J_\alpha(\mathbf{P}', \mathbf{P})\right] \tag{2.79}$$

where

$$\Sigma_t = \frac{m}{P} \bar{w}$$

$$\Sigma_s(\mathbf{P}') \mathscr{F}(\mathbf{P}' \to \mathbf{P}) = \frac{m}{P'} w_s(\mathbf{P}' \to \mathbf{P})$$

$$\Sigma_f(\mathbf{P}') B^J_\alpha(\mathbf{P}', \mathbf{P}) = \frac{m}{P'} \sum_{\{\mathbf{K}_i\}_J} \overline{w}^F_{\mathbf{K}' \to \{\mathbf{K}_i\}_J} Q^\alpha_{\mathbf{K}\{\mathbf{K}_i\}_J} \tag{2.80}$$

The frequency B^J_α is introduced such that $B^J_\alpha(\mathbf{P}', \mathbf{P})\, \mathrm{d}^3P$ is the probability that a fission induced by a neutron at $\mathbf{P}'$ will produce exactly J neutrons, α of which have momenta in d^3P about $\mathbf{P}$.

E. Effect of an External Field

In closing this chapter we consider briefly the effect of a time-independent external field, $\phi_{\text{ext}}(\mathbf{x})$.* The Hamiltonian H is now modified by the addition of V_{ext},

$$V_{\text{ext}} = \int \mathrm{d}^3x\, \psi_j^+(\mathbf{x})\, \psi_j(\mathbf{x})\, \phi_{\text{ext}}(\mathbf{x}) \tag{2.81}$$

In the decomposition of H we shall group V_{ext} with T' so that the effect of the external field appears only in the commutator, $[V_{\text{ext}}, \varrho_1(\mathbf{X}, \mathbf{K})]$.

It will be convenient to evaluate the present commutator in a manner somewhat different from the way in which the streaming term is derived. We note that an integral representation of the number operator ϱ_1 is

$$\varrho_1(\mathbf{X}, \mathbf{K}) = \int \mathrm{d}^3x'\, \mathrm{d}^3x\, \varphi^*(\mathbf{X}, \mathbf{K}, \mathbf{x})\, \varphi(\mathbf{X}, \mathbf{K}, \mathbf{x}')\, \psi_j^+(\mathbf{x})\, \psi_j(\mathbf{x}') \tag{2.82}$$

* An example of $\phi_{\text{ext}}(\mathbf{x})$ is the gravitational field.

Then from Eqs. 2.81 and 2.82

$$[V_{\text{ext}}, \varrho_1] = \int d^3x\, d^3x'\, d^3x''\, \varphi^*(\mathbf{X}, \mathbf{K}, \mathbf{x})\, \varphi(\mathbf{X}, \mathbf{K}, \mathbf{x}')\, \phi_{\text{ext}}(\mathbf{x}'') \times$$
$$\times\, [\psi_{j'}^+(\mathbf{x}'')\, \psi_{j'}(\mathbf{x}''), \psi_j^+(\mathbf{x})\, \psi_j(\mathbf{x}')]$$
$$= \int d^3x\, d^3x'\, \varphi^*(\mathbf{X}, \mathbf{K}, \mathbf{x})\, \varphi(\mathbf{X}, \mathbf{K}, x')\, \psi_j^+(\mathbf{x})\, \psi_j(\mathbf{x}') \times$$
$$\times\, [\phi_{\text{ext}}(\mathbf{x}') - \phi_{\text{ext}}(\mathbf{x})] \tag{2.83}$$

Because of the cell functions the x and x' integrals only extend over the cell centered at $\mathbf{X}$. If now ϕ_{ext} is a slowly-varying function over a distance of order L, it may be approximately represented as

$$\phi_{\text{ext}}(\mathbf{x}) \approx \phi_{\text{ext}}(\mathbf{X}) + (x_j - X_j)\left(\frac{\partial \phi_{\text{ext}}}{\partial x_j}\right)_{x_j = X_j} \tag{2.84}$$

Entering this expression into Eq. 2.83 we find

$$[V_{\text{ext}}, \varrho_1] \approx L^{-3}\left(\frac{\partial \phi_{\text{ext}}}{\partial x_k}\right)_{x_k = X_k} \int d^3x\, d^3x'\, E(\mathbf{X}, \mathbf{x}')\, E(\mathbf{X}, \mathbf{x}) \times$$
$$\times\, (x_k' - x_k)\, \psi_j^+(\mathbf{x})\, \psi_j(\mathbf{x}')\, e^{-i\mathbf{K}\cdot(\mathbf{x}'-\mathbf{x})}$$
$$= i\left(\frac{\partial \phi_{\text{ext}}}{\partial x_j}\right)_{x_j = X_j} \frac{\partial \varrho(\mathbf{X}, \mathbf{K})}{\partial \mathbf{K}_j} \tag{2.85}$$

The effect of an external field on the transport equation, Eq. 2.56, is thus described by the additional term

$$-\frac{i}{\hbar L^3}\, \text{Tr}\, [V_{\text{ext}}, \varrho_1(\mathbf{X}, \mathbf{K})]\, D(t) \approx \left(\frac{\partial \phi_{\text{ext}}}{\partial x_j}\right)_{x_j = X_j} \frac{\partial F(\mathbf{X}, \mathbf{K}, t)}{\partial P_j} \tag{2.86}$$

which, like the streaming term, is a familiar result in fine-grained theories. Note that in the present instance coarse-grained momentum is treated like a continuous variable. This procedure is acceptable so long as $K_j L \gg 2\pi$, which then represents a lower limit in the choice of cell size. A corresponding upper limit is determined by the truncated series expression of ϕ_{ext}, Eq. 2.84.

References

1. S. Ono, *Prog. Theor. Phys.* (Japan), **12**: 113 (1954).
2. L. I. Schiff, *Quantum Mechanics*, McGraw-Hill Book Co., Inc., New York, 1955.
3. P. A. M. Dirac, *The Principles of Quantum Mechanics*, Clarendon Press, Oxford, 1958.

4. J. von Neumann, *Mathematical Foundations of Quantum Mechanics*, Princeton University Press, Princeton, N.J., 1955.
5. R.C. Tolman, *The Principles of Statistical Mechanics*, Oxford University Press, Oxford, 1948.
6. E.W. Gibbs, *Elementary Principles in Statistical Mechanics*, Yale University Press, New Haven, 1902.
6a. N.G. van Kampen, *Proceedings of the International Symposium on Transport Processes in Statistical Mechanics*, Interscience Publishers, New York, 1958.
6b. U. Fano, *Revs. Mod. Phys.*, **29**: 74 (1957).
7. H. Mori, *Prog. Theor. Phys.* (Japan), **9**: 473 (1953).
8. A.W. Saenz, *Phys. Rev.*, **105**: 546 (1957).
9. F. Bloch, *Zeits. f. Phys.*, **52**: 555 (1928); **59**: 208 (1930).
10. E.A. Uehling and G.E. Uhlenbeck, *Phys. Rev.*, **43**: 552 (1933).
11. E.P. Wigner, *Phys. Rev.*, **40**: 749 (1932); J.H. Irving and R.W. Zwanzig, *J. Chem. Phys.*, **19**: 1173 (1951).
12. R.K. Osborn, *Phys. Rev.*, **130**: 2142 (1963).
13. R.K. Osborn and E.H. Klevans, *Ann. Phys.*, **15**: 105 (1961); E.H. Klevans, Thesis, University of Michigan, Ann Arbor, Michigan, 1962.
14. S. Chandrasekhar, *Radiative Transfer*, Dover Publications, Inc., New York, 1960.
15. H. Goldstein, *Fundamental Aspects of Reactor Shielding*, Addison-Wesley, Reading, Mass., 1959.
16. See, for example, G.W. Uhlenbeck and G.W. Ford, *Lectures in Statistical Mechanics*, American Mathematical Society, Providence, R.I., 1963.
17. L. van Hove, *Physica*, **21**: 517 (1955); J. Luttinger and W. Kohn, *Phys. Rev.*, **109**: 1892 (1958).
18. T.D. Schultz, "Quantum Field Theory and the Many-Body Problem", Space Technology Laboratory Report STL/TR-60-0000-GR-332, 1960; L.P. Kadanoff and G. Baym, *Quantum Statistical Mechanics*, Benjamin, New York, 1962; D. Pines, *The Many-Body Problem*, Benjamin, New York, 1962.

III

Neutron-nuclear Interactions: Mainly Nuclear Considerations

In this chapter we will undertake an investigation into the effects of neutron-nuclear collisions upon the balance relation (2.56). Many kinds of nuclear reactions may be initiated by such collisions. However, we shall concentrate our attention on only a few of them. There are at least two reasons for this restriction. In the first place there are only a few such reactions that can be dealt with at all adequately by the rather elementary analytical techniques that we envisage here. In the second place our main emphasis is on an illustrative investigation of the basis of the theory of the distribution of relatively low-energy neutrons. Consequently, fission, radiative capture, and elastic scattering are probably an adequate sample of representative and significant interactions.

There are two types of effects that must be taken into account in the description of a collision process—the specifically nuclear effects and the effects of the macroscopic medium. The former depend upon nuclear forces, while the latter depend upon the non-nuclear interactions of the nucleus with its surroundings in the system. Because of their importance, it is essential that the present discussion of the transition probability per unit time, $W_{n'n}$, be made sufficiently general to include both nuclear and medium effects.

The specifically nuclear effects can be treated by means of the steady-state theory of nuclear reactions as, say, presented by Blatt and Weisskopf[1] and reviewed by Lane.[2] However, it is not clear that chemical binding effects can be conveniently grafted onto this elegant and rigorous theory of binary nuclear reactions. At the opposite extreme we have straightforward perturbation methods for a successive approximation evaluation of $W_{n'n}$. Although such an approach enables easy incorporation of the medium effects, this extreme must also be avoided since it appears that only potential scattering can be readily and usefully treated in this manner. For indirect processes, many of which contribute significantly to (2.56), conventional perturbation methods are therefore not adequate.

As a compromise, we will follow an approach originally developed by Heitler[3] in the study of photon interactions with matter. The theory, sometimes known as damping theory,[3–5] is sufficiently elementary so that both medium and nuclear effects can be considered and is, at the same time, sufficiently sophisticated to allow a useful exposition of the essential features of both types of phenomena.

From the point of view of the neutron transport equation, some of the collisions of importance are those that result in capture, elastic and inelastic scattering, and fission. We do not regard as an essential part of our purpose the detailed investigations of the specifically nuclear effects of these reactions since very complete and thorough discussions are available in the literature.* For this reason many aspects of the following calculations will not be explored as fully as possible. Moreover, not all the reactions are treated with equal emphasis. It will be seen that considerable detail is presented in the study of radiative capture and elastic scattering, whereas the discussions of inelastic scattering and fission processes are brief and, at best, descriptive. This by no means is intended to imply the relative importance of the reactions in the transport equation in general, although there are special cases in which the effects of a given reaction or reactions are suppressed. What we attempt here, in essence, is to illustrate another approach to nuclear reaction theory that is capable of producing, at least qualitatively, the conventional results and also allows a systematic treatment of the external degrees of freedom of the nucleus.†

A. Formal Development of the Transition Probability

The task of evaluating the transition matrix $W_{n'n}$ is essentially that of determining the off-diagonal matrix elements of the "temporal evolution" operator $U'(\tau)$. Thus far, the representation in which the matrix elements are to be calculated is only required to diagonalize the neutron

---

* See reference 1 for a general discussion of theory of nuclear reactions. For those aspects of particular interest in reactor physics see Weinberg and Wigner.[6]

† The present approach has also been employed in recent studies of photon transport in dispersive[7] media, and of line shape theory.[8,9]

number operator; otheiwise it is unspecified. With the diagonalization of ϱ_1, we observe that $\mathscr{E}$, the kinetic energy of neutrons within cells, also becomes diagonal.

To develop a general expression for $U'_{n'n}(\tau)$ some consideration must be given to other degrees of freedom of the system. For the moment they need only to be introduced formally, detailed discussion being necessary only when a specific reaction is to be investigated. We therefore further specify that the above representation also diagonalize the operator H_s, i.e.

$$(\mathscr{E} + H_s)\,|n\rangle = \varepsilon_n\,|n\rangle \tag{3.1}$$

The states $|n\rangle$ are assumed to form an orthonormal and complete set. Although H_s describes the entire system exclusive of neutrons, the only part of it that will require our subsequent attention, in view of the reactions of interest, is that relevant to the description of nuclei (and of photons in the case of resonance capture).

The operator $U'(\tau)$, as given by (2.47), has as its Laplace transform

$$G(z) = \int_0^\infty \frac{d\tau}{\hbar}\, U'(\tau)\, e^{-\tau z/\hbar} = (z + iH')^{-1} \tag{3.2}$$

where $H' = \mathscr{E} + H_s + V$. The form of this operator is particularly suitable for developing approximations. As will be seen, the present calculation provides an approximate expression for the off-diagonal matrix elements of $G(z)$. Once $G_{n'n}(z)$ is known, the inversion then gives

$$U'_{n'n}(\tau) = \frac{1}{2\pi i}\int_{\gamma - i\infty}^{\gamma + i\infty} dz\, G_{n'n}(z)\, e^{\tau z/\hbar} \tag{3.3}$$

From Eq. 3.2 we have the matrix equation

$$(z + i\varepsilon_n)\, G_{nn'} + i \sum_{n''} V_{nn''} G_{n''n'} = \delta_{nn'} \tag{3.4}$$

It will be convenient to treat the diagonal and nondiagonal parts of G separately. For this purpose we introduce an operator Q such that

$$G_{nn'} = G_{nn}\, Q_{nn'}\, G_{n'n'} \tag{3.5}$$

for $n \neq n'$.

The diagonal elements of G then satisfy the formal relation

$$G_{nn}(z) = \left(z + i\varepsilon_n + \frac{i\hbar}{2}\gamma_n\right)^{-1} \tag{3.6}$$

where

$$\frac{\hbar}{2}\gamma_n(z) = V_{nn} + \sum_{n' \neq n} V_{nn'}\, G_{n'n'}\, Q_{n'n} \tag{3.7}$$

An essential step in the development is the determination of $Q_{nn'}$. From Eq. 3.4 we obtain

$$[z + i(\varepsilon_n + V_{nn})]\, G_{nn}\, Q_{nn'} + iV_{nn'} + i \sum_{n'' \neq n,n'} V_{nn''}\, G_{n''n''}\, Q_{n''n'} = 0 \tag{3.8}$$

or

$$\begin{aligned} Q_{nn'} = & -iV_{nn'} - i \sum_{n'' \neq n,n'} V_{nn''}\, G_{n''n''}\, Q_{n''n'} \\ & + i \sum_{n'' \neq n} V_{nn''}\, G_{n''n''}\, Q_{n''n}\, G_{nn}\, Q_{nn'} \end{aligned} \tag{3.9}$$

A useful, approximate solution to this equation can be obtained by expanding Q as a power series in V, *and ignoring the dependence of G on V*. This is readily accomplished by writing

$$\begin{aligned} V &\to \lambda V \\ Q &\to \sum_{\sigma=0}^{\infty} \lambda^{\sigma+1}\, Q^{(\sigma)} \end{aligned} \tag{3.10}$$

and considering λ as a bookkeeping parameter ultimately to be evaluated at the unit point. We find, to second order in λ

$$Q_{nn'}(z) = -iV_{nn'} - \sum_{n'' \neq n,n'} \frac{V_{nn''}\, V_{n''n'}}{z + i\varepsilon_{n''} + \frac{i\hbar}{2}\gamma_{n''}} \tag{3.11}$$

and

$$\frac{i\hbar}{2}\gamma_n(z) = iV_{nn} + \sum_{n' \neq n} \frac{|V_{nn'}|^2}{z + i\varepsilon_{n'}} \tag{3.12}$$

With $\gamma_n(z)$ given by Eq. 3.12, the diagonal elements of G are now explicitly determined by Eq. 3.6. The higher-order terms which have been ignored can be investigated. But we anticipate that the predominant features of the reactions of interest are usefully described by the two terms in Eq. 3.11. To this order of approximation, the off-diagonal elements of $U'(\tau)$ are given as

$$\begin{aligned} U'_{nn'}(\tau) \approx & \frac{1}{2\pi i}\int_{\gamma - i\infty}^{\gamma + i\infty} dz\, \bar{A}_n(z)\, \bar{A}_n(z)\, e^{z\tau/\hbar} \times \\ & \times \Big[-iV_{nn'} - \sum_{n'' \neq n,n'} V_{nn''}\, V_{n''n'}\, A_{n''}(z)\Big] \end{aligned} \tag{3.13}$$

where

$$\bar{A}_n(z) = \left[z + i\varepsilon_n + \frac{i\hbar}{2}\gamma_n(z)\right]^{-1} \tag{3.14}$$

The quantity $\gamma_n(z)$ is the width and shift function for the energy level corresponding to the nth eigenstate of the system. It will be shown that when evaluated at $z = -i\varepsilon_n$ the real part gives the shift of the unperturbed energy level due to interactions, while its imaginary part describes the width and hence the finite lifetime of the state $|n\rangle$. The only reason that we cannot neglect these quantities completely in the present problem is that many of the neutron-nuclear reactions we are concerned with proceed via excited states which are known to be significantly broadened. At the same time, in the systems of predominant concern here, it is most probable that the interacting nuclei are initially in their internal ground states. If also we consider only time intervals (τ) long compared to the lifetimes of the intermediate states,* the final states can also be taken as ground states so far as the specifically nuclear degrees of freedom are concerned. Hence for our purposes the only states whose widths and shifts will have appreciable influence on the collisions are the intermediate states. We will accordingly ignore the width and shift functions in $\bar{A}_n(z)$ and $\bar{A}_{n'}(z)$ in Eq. 3.14, thus

$$U'_{nn'}(\tau) \approx \frac{1}{2\pi i}\int_{\gamma - i\infty}^{\gamma + i\infty} dz\, \bar{B}_n(z)\, \bar{B}_{n'}(z)\, e^{z\tau/\hbar} \times$$

$$\times \left[-iV_{nn'} - \sum_{n''\neq n,n'} V_{nn''}\, V_{n''n'}\, \bar{A}_{n''}(z)\right] \tag{3.15}$$

where

$$\bar{B}_n(z) = (z + i\varepsilon_n)^{-1} \tag{3.16}$$

In Eq. 3.15 the two terms represent the effects of direct and two-stage (compound nucleus) processes respectively. It is anticipated that the first term will suffice to describe potential scattering, whereas the second term will lead to a description of resonance reactions (including fission). For elastic scattering, both terms must be considered simultaneously thus enabling an examination of potential scattering, elastic resonance scattering, and the interference between them.

The evaluation of the first integral in Eq. 3.15 is a simple matter, and

* This provides a qualitative lower limit for τ.

we find

$$\frac{1}{2\pi i}\int_{\gamma-i\infty}^{\gamma+i\infty} dz\, \bar{B}_n(z)\, \bar{B}_{n'}(z)\, e^{z\tau/\hbar} = -\frac{e^{-i\omega_n\tau}}{i\hbar\omega_{nn'}}(1 - e^{i\omega_{nn'}\tau}) \quad (3.17)$$

where $\hbar\omega_n = \varepsilon_n$, and $\hbar\omega_{nn'} = \varepsilon_n = \varepsilon_{n'}$. The second integral can be expressed as a convolution[10]

$$h(\tau) = \frac{1}{2\pi i}\int_{\gamma-i\infty}^{\gamma+i\infty} dz\, \bar{B}_n(z)\, \bar{B}_{n'}(z)\, \bar{A}_{n''}(z)\, e^{z\tau/\hbar}$$

$$= \hbar^{-2}\int_0^{\tau} d\tau' A_{n''}(\tau - \tau')\int_0^{\tau'} d\tau''\, B_{n'}(\tau - \tau'')\, B_n(\tau'') \quad (3.18)$$

where

$$B_n(\tau) = e^{-i\tau\varepsilon_n/\hbar} \quad (3.19)$$

and $A_n(\tau)$ is the inverse Laplace transform of $\bar{A}_n(z)$. A discussion of the function $A_n(\tau)$ has been given by Akcasu.[9] We shall adopt here a slightly different approach. Since

$$\bar{B}_n(z)\, \bar{B}_n(z) = \frac{1}{i(\varepsilon_{n'} - \varepsilon_n)}\, [\bar{B}_n(z) - \bar{B}_{n'}(z)]$$

we then have

$$h(\tau) = M_n(\tau) - M_{n'}(\tau) \quad (3.20)$$

with

$$M_n(\tau) = \frac{1}{2\pi i}\int_{\gamma-i\infty}^{\gamma+i\infty} dz\, \bar{A}_{n''}(z)\, \bar{B}_n(z)\, e^{z\tau/\hbar}$$

$$= \frac{1}{2\pi i}\int_{\gamma-i\infty}^{\gamma+i\infty} dz \int_0^{\infty}\frac{d\tau'}{\hbar}\, A_{n''}(\tau)\, \bar{B}_n(z)\, e^{z(\tau-\tau')/\hbar}$$

$$= \int_0^{\infty}\frac{d\tau'}{\hbar}\, A_{n''}(\tau)\, B_n(\tau - \tau') \quad (3.21)$$

The function $B_n(\tau - \tau')$ is given by (3.19) and vanishes for $\tau' > \tau$, so

$$M_n(\tau) = e^{-i\varepsilon_n\tau/\hbar}\int_0^{\tau}\frac{d\tau'}{\hbar}\, A_{n''}(\tau)\, e^{-i\varepsilon_n\tau/\hbar} \quad (3.22)$$

Next we would like to extend the upper limit of integration to infinity. This procedure is justified if $A_{n''}(\tau')$ is negligible in the region $\tau' > \tau$, and such is the case if in $\bar{A}_{n''}(z)$ the quantity $\gamma_{n''}(z)$ is essentially independent of z and has an imaginary part much larger than $1/\tau$. As we shall show in the following, Im $(\gamma_{n''})$ is a measure of the reciprocal of the lifetime of the intermediate state, $|n''\rangle$. Hence, by writing

$$M_n(\tau) = \bar{A}_{n''}(-i\varepsilon_n)\, e^{-i\varepsilon_n\tau/\hbar} \tag{3.23}$$

it is implied that one is only looking at that part of $U'_{nn'}$ which describes the completed transition from initial state, $|n\rangle$, to final state, $|n'\rangle$. We now obtain

$$h(\tau) = \frac{i\, e^{-i\varepsilon_n\tau/\hbar}}{\varepsilon_n - \varepsilon_{n'}} [\bar{A}_{n''}(-i\varepsilon_n) - \bar{A}_{n''}(-i\varepsilon_{n''})\, e^{i(\varepsilon_n - \varepsilon_{n'})\tau/\hbar}] \tag{3.24}$$

For a fixed ε_n the most important contribution arises when $\varepsilon_{n'} = \varepsilon_n$,* thus

$$h(\tau) = i(\hbar\omega_{nn'})^{-1} A_{n''}(-i\varepsilon_n)\,(1 - e^{i\omega_{nn'}\tau})\, e^{-i\omega_n\tau} \tag{3.25}$$

Combining this result with (3.17) we find

$$U'_{nn'} \approx \frac{1}{\hbar}\left[V_{nn'} - \sum_{n''\neq n,n'} \frac{V_{nn''}\, V_{n''n'}}{\varepsilon_{n''} - \varepsilon_n + (\hbar/2)\,\gamma_{n''}(-i\varepsilon_n)}\right] \times$$
$$\times\ \frac{(1 - e^{i\omega_{nn'}\tau})\, e^{-i\omega_{nn'}\tau}}{\hbar\, \omega^2_{nn'}} \tag{3.26}$$

Our result shows that the function $\gamma_{n''}$ is to be evaluated at $-i\varepsilon_n$. The behavior of this function along the imaginary axis is readily examined by comparing the boundary values of $\gamma_{n''}(z)$ as the axis is approached from both sides of the complex plane. One finds

$$\lim_{x\to 0+} \frac{i\hbar}{2}\, \gamma_n(x + iy) = is_n(y) + \frac{1}{2}\, \Gamma_n(y) \tag{3.27}$$

$$\lim_{x\to 0-} \frac{i\hbar}{2}\, \gamma_n(-x + iy) = is_n(y) - \frac{1}{2}\, \Gamma_n(y) \tag{3.28}$$

---

* This is merely to say that we anticipate the conservation of energy. To see that $h(\tau)$, for large τ, is sharply peaked about $\omega_{nn'} = 0$, consider

$$h(\tau) = i\, e^{-i\omega_n\tau}(\hbar\omega_{nn'})^{-1} \times$$
$$\times \left[A_{n''}(-i\varepsilon_n)\,(1 - e^{i\omega_{nn'}\tau}) - i\omega_{nn'}\left(i\frac{\partial A_{n''}}{\partial\varepsilon_n}\right) e^{i\omega_{nn'}\tau} + O(\omega^2_{nn'})\right]$$
$$\xrightarrow[\omega_{nn'}\to 0]{} \frac{i}{\hbar}\, e^{-i\omega_n\tau}\left[-iA_{n''}(-i\varepsilon_n)\,\tau - i\left(i\frac{\partial A_{n''}}{\partial\varepsilon_n}\right) + O(\omega_{nn'})\right]$$

so the first term dominates.

where

$$s_n(y) = V_{nn} - \sum_{n' \neq n} |V_{nn'}|^2 \mathscr{P} \frac{1}{y + \varepsilon_{n'}} \tag{3.29}$$

$$\Gamma_n(y) = \pi \sum_{n' \neq n} |V_{nn'}|^2 \, \delta(y + \varepsilon_{n'}) \tag{3.30}$$

In obtaining these expressions from (3.12) we have made use of the relation*

$$\lim_{x \to 0} \frac{1}{x + iy} = -i \mathscr{P} \frac{1}{y} + \pi\delta(y) \tag{3.31}$$

where $\mathscr{P}(1/x)$ is the principal value of $1/x$. Thus in crossing the imaginary axis the value of $\gamma_{n'}$ changes by an amount $2\Gamma_{n''}(y)$ which vanishes everywhere except at $y = -\varepsilon_{n0}$, $n_0 \neq n''$. Hence $\gamma_{n''}(z)$ has branch points at $z = -i\varepsilon_{n_0}$, $n_0 \neq n''$.

The transition matrix is obtained once $U'_{nn'}$ is known, i.e.

$$W_{nn'} \approx \frac{2\pi}{\hbar} \left| V_{nn'} - \sum_{n'' \neq n, n'} \frac{V_{nn''} V_{n''n'}}{\varepsilon_{n''} - \varepsilon_n + \frac{\hbar}{2} \gamma_{n''}(-i\varepsilon_n)} \right|^2 \left(\frac{1 - \cos \omega_{nn'}\tau}{\pi\tau\omega_{nn'}^2} \right) \tag{3.32}$$

For sufficiently large τ, i.e., $\omega_{nn'}\tau \gg 1$, the last factor in Eq. 3.32 is a sharply peaked function about $\omega_{nn'} = 0$,† and can be replaced by a delta function. The quantity $\omega_{nn'}\tau$ will always be substantially greater than unity if the width of the intermediate state is small compared to its energy above ground state, i.e., if $\hbar\omega_{nn'} > \gamma_n$. If such is not the case, the notion of the intermediate state becomes fuzzy and so does the concept of the transition probability per unit time.

We have obtained a useful, though approximate, expression for $W_{nn'}$. It is convenient to exhibit this general result as

$$W_{n'n} = \delta(\varepsilon_n - \varepsilon_{n'}) \, R_{n'n} \tag{3.33}$$

$$R_{n'n} = \frac{2\pi}{\hbar} \left| V_{n'n} - \sum_{n'' \neq n, n'} \frac{V_{n'n''} V_{n''n}}{\varepsilon_{n''} - \varepsilon_n + \frac{\hbar}{2} \gamma_{n''}(-i\varepsilon_n)} \right|^2 \tag{3.34}$$

The transition probability is independent of τ as one might expect in

---

* See Heitler, ref. 3, p. 70.

† See Schiff, reference 11, p. 198.

the present situation. To develop explicit cross-section formulas for the various reactions, it is now necessary to consider in more detail the states $\{|n\rangle\}$ and the matrix elements of the interaction, V.

The eigenstates $\{|n\rangle\}$ were introduced in Eq. 3.1 simply as a diagonalizing representation for the kinetic energy of free neutrons in cells and for the total energy of the "system" with which the neutrons interacted according to a potential V. We shall regard the "system" as an assembly of electrons, photons, and nuclei of various kinds. In the present study we will ignore the interactions between neutrons and the electrons[12] on the ground that they have little effect on neutron transport.* Consequently, electronic coordinates appear only in H_s. We will also ignore the photon-neutron coupling; however, photon coordinates will appear in both H_s and V because it is convenient to incorporate the energy of free photons in the former and the interaction of photons with nuclei in the latter. It is necessary to take explicit account of the photons only for the description of radiative capture; for the other processes to be considered here the presence of photons has little influence on the cross sections.

Following the above remarks, we exhibit the eigenstate $|n\rangle$ as a product of eigenstates appropriate to each kind of particle,

$$|n\rangle = |N_{\mathbf{X},\mathbf{K},s}\rangle\, |N_{\mathbf{X},\varkappa,\lambda}\rangle\, |N_{A,\alpha,k}\rangle \tag{3.35}$$

The eigenstates for the neutrons and the labels that characterize them were introduced in Chapter II. There it was mentioned that a neutron state $|n\rangle$, denoted here as $|N_{\mathbf{X},\mathbf{K},s}\rangle$, is completely specified by a set of occupation numbers for all spin and momentum states and cell labels. From Eq. 2.18,

$$|N_{\mathbf{X},\mathbf{K},s}\rangle = |N(\mathbf{X}_1, \mathbf{K}_1, s_1)\, N(\mathbf{X}_1, \mathbf{K}_1, s_2) \ldots N(\mathbf{X}_j, \mathbf{K}_j, s_j) \ldots\rangle \tag{3.36}$$

It is appropriate to treat the photons also by the field formalism. Then the photon eigenstates will be specified by a set of occupation numbers for all polarization and momentum states and cell labels,

$$|N_{\mathbf{X},\varkappa,\lambda}\rangle = |N(\mathbf{X}_1, \boldsymbol{\kappa}_1, \lambda_1)\, N(\mathbf{X}_1, \boldsymbol{\kappa}_1, \lambda_2) \ldots N(\mathbf{X}_j, \boldsymbol{\kappa}_j, \lambda_j) \ldots\rangle \tag{3.37}$$

where $N(\mathbf{X}, \boldsymbol{\kappa}, \lambda)$ is the number of photons in cell $\mathbf{X}$ with momentum $\hbar\boldsymbol{\kappa}$ and polarization λ. Since photons are bosons, this number can be any positive integer or zero.

---

* The inclusion of neutron-electron interactions entails no difficulty in principle.

The eigenstates for the entire collection of interacting nuclei and electrons are less easily described and more cumbersome to deal with. In the first place, like the neutrons and the photons the nuclei are not conserved, so that one is tempted toward a field formalism for their description. But on the other hand, the nuclei may well be localized, as atoms bound in crystals, thus making the application of field theory awkward if not obscure. If, in fact, the nuclei (atoms or molecules) are in gas phase, then their treatment in analogy to that of the neutrons and photons would be quite appropriate. However, for the general discussion (more applicable to solids and liquids) we will make use of eigenvectors whose components themselves are many-particle configuration-space wave functions describing definite numbers of nuclei of definite kinds. Different components would then describe different numbers of nuclei of definite kinds. These eigenvectors will be presumed to be orthonormal, and it will be further presumed that V has some non-vanishing off-diagonal matrix elements with respect to these representations. As a notation we will write

$$|N_{A,\alpha,k}\rangle = |N(A_1, \alpha_1, k_1)\, N(A_1, \alpha_1, k_2) \ldots N(A_j, \alpha_j, k_j) \ldots\rangle \quad (3.38)$$

to represent a nuclear state with $N(A_1, \alpha_1, k_1)$ nuclei of kind A_1 with internal and external states specified by labels α_1 and k_1 respectively, etc. It is important to keep in mind that the components of these vectors are not functions in occupation number space, but rather in ordinary configuration and spin space.

We will treat the various interactions separately. Following the approach outlined in Chapter II we decompose all interactions into classes according to the relative number of particles of a given kind in the states $|n\rangle$ and $|n'\rangle$. This will be seen to be a natural way of classifying the different *binary* neutron-nuclear reactions. Scattering reactions, both potential and resonance scattering, are characterized by the same total number of neutrons in the final state as in the initial state. This is true for both elastic and inelastic events, although inelastic scattering really belongs to a subclass in which the number of photons in the final state differs from that in the initial state.* If the neutron and the photon(s) are emitted separately in an inelastic scattering process, such an event will require a description that allows at least two intermediate states.

---

* We continue to treat the nuclei in both initial and final states as in their internal ground states.

Since the present treatment is restricted to only one intermediate state, our discussion of scattering will initially be limited to elastic processes. Later, we will assume that the approximation in which the compound nucleus decays to ground state by a simultaneous emission of neutron and photon is adequate for treating inelastic scattering. Radiative capture reactions, as well as all other neutron capture processes which are followed by a decay to ground, are distinguished by one less neutron in the final state than in the initial state. Finally fission is a reaction in which the neutron number in the final state may be increased by one or more with respect to that of the initial state. Thus in the following we shall consider radiative capture, scattering and fission reactions. Though these hardly exhaust all the interesting possibilities, they are the main processes that significantly influence neutron transport in many reactor situations.

As an initial step in the reduction of collision terms in Eq. 2.56 we rewrite Eq. 2.56 as

$$\left(\frac{\partial}{\partial t} + \frac{\hbar}{m} K_j \frac{\partial}{\partial X_j}\right) F(\mathbf{X}, \mathbf{K}, t)$$

$$= -V^{-1} \sum_{nn's} W^{c}_{n'n} D_{nn} + V^{-1} \sum_{nn's} W^{sG}_{n'n} D_{nn} - V^{-1} \sum_{nn's} W^{sL}_{n'n} D_{nn}$$

$$+ V^{-1} \sum_{nn's} [N'(\mathbf{X}, \mathbf{K}, s) - N(\mathbf{X}, \mathbf{K}, s)]\, W^{F}_{n'n} D_{nn} \qquad (3.39)$$

where we decompose the n' sum for a given n into sums corresponding to the different types of $W_{n'n}$. The terms proportional to $W^{sG}_{n'n}$ are all those for which the final states contain the same total number as the initial and for which $N'(\mathbf{X}, \mathbf{K}, s) = N(\mathbf{X}, \mathbf{K}, s) + 1$. They are therefore the scattering gain contributions to the balance relation in the binary collision approximation. Analogously, the terms containing $W^{sL}_{n'n}$ constitute the scattering loss contribution.* The terms containing $W^{c}_{n'n}$ are all those (except fission) for which the total neutron number in the final state is one less than in the initial state and for which $N'(\mathbf{X}, \mathbf{K}, s) = N(\mathbf{X}, \mathbf{K}, s) - 1$. These represent the effect of neutron capture reactions. The companion terms representing neutron gain by emission from excited nuclei have been neglected in writing Eq. 3.39.† Finally the

* The scattering gain and loss terms will consist of both elastic and inelastic contributions.

† This is not justified if, say, the concentration of photo neutrons in the system is appreciable.

terms containing $W_{n'n}^{F}$ are to represent the fission contribution in which an arbitrary increase in the number of neutrons is allowed. A number of other binary interactions could be included in Eq. 3.39, however, they are of more special interest* and need not be considered in a general discussion of collision effects in neutron transport.

The following sections in this chapter will be devoted to a study of the specifically nuclear aspects of the various transition probabilities indicated in Eq. 3.39. When reduced, the collision terms will have the same form as those discussed in the previous chapter, but in the present instance explicit expressions for the reduced transition probabilities will be derived. In the next chapter the influence of macroscopic medium effects will be investigated in some detail.

B. Radiative Capture†

The radiative capture reaction (n, γ) is not the simplest reaction considered in the present work. It is generally viewed as a two-stage process involving the passage through an intermediate state. Consequently, a more complicated description is required than that for the direct process of elastic potential scattering. However, a general treatment of elastic scattering must also include considerations of resonant scattering, a process of the same order of complexity as radiative capture. Thus we shall first examine the (n, γ) reaction and will make use of certain features of the resonance process in general in later discussions of elastic scattering.

The (n, γ) reaction is schematically represented by

$$n + {}_zX^A \rightarrow {}_zX^{A+1*} \rightarrow \gamma + {}_zX^{A+1} \tag{3.40}$$

where we assume that the neutron interacts with the nucleus to form a compound nucleus which then decays directly to its ground state via the emission of a photon. The transition probability $W_{n'n}^{c}$ associated

---

* For example, the $(n, 2n)$ reaction in beryllium.

† Other capture reactions such as (n, p) and (n, α) will not be considered here. Their contributions to the transport equation can usually be ignored (see, for example, reference 6, p. 51).

The reader may see Dresner[13] for a thorough investigation of the effects of resonance absorption of neutrons in nuclear reactors; see also the work of Nordheim[14] and a review by Sampson and Chernick.[15]

with this process is given formally by Eqs. 3.33 and 3.34. The potential V describes both neutron-nuclear and nuclei-electromagnetic interactions. These interactions are presumed to be separable in the sense that

$$V = V^{\gamma} + V^{N} \tag{3.41}$$

where V^N involves the specifically nuclear forces and is that part of V that causes the transition from initial to intermediate state (neutron absorption), and V^{γ} is the electromagnetic part that causes the transition from intermediate to final state (photon emission). A consequence of this separation is that V will have no nonvanishing matrix elements in which *both* the neutron and the photon numbers are changed.* The reduced transition matrix thus becomes

$$R^{c}_{n'n} = \frac{2\pi}{\hbar}\left|\sum_{n''\neq n,n'} \frac{V^{\gamma}_{n'n''}V^{N}_{n''n}}{\varepsilon_{n''} - \varepsilon_{n} + \frac{\hbar}{2}\gamma_{n''}(-i\varepsilon_{n})}\right|^{2} \tag{3.42}$$

It is now appropriate to obtain a more explicit expression for the width and shift function γ. We recall from Eq. 3.12 the expression

$$\frac{\hbar}{2}\gamma_{n''}(-i\varepsilon_{n}) = V_{n''n''} + \sum_{m\neq n''} \frac{|V_{n''m}|^{2}}{\varepsilon_{n} - \varepsilon_{m}} \tag{3.43}$$

The m sum is to be regarded as a summation over all possible sets of neutron and photon occupation numbers, and over all states of the nuclei. This sum can be decomposed into contributions arising from those states like the initial state, those like the final state, those like the intermediate state, and all other states for which $V_{n''m}$ does not vanish. The contributions from the last class of states are not considered here and will henceforth be neglected.† Note that by two alike states we mean that the number of any given kind of particles is the same in both states, and nothing is to be inferred about their respective momentum and spatial distributions.

The contribution to the m sum in Eq. 3.43 from states like the initial

* In treating inelastic scattering we will find it necessary to violate this condition. This is, however, because we insist on using Eq. 3.34 to describe what is essentially a three-stage process.

† An example of such a contribution is that which describes the decay of the intermediate state by proton or alpha particle emission as in (n, p) or (n, α) reactions.

state may be written in the form

$$\sum_{k_j, \mathbf{K}_j, s_j} \frac{|V^N_{k''\alpha'', k_j K_j s_j}|^2 \, [1 - N(\mathbf{X}, \mathbf{K}_j, s_j)]}{E^A_k + E_K - E_{K_j} - E^A_{k_j}} \tag{3.44}$$

where we designate as E_K and E^A_k respectively the kinetic energy of a neutron with momentum $\hbar\mathbf{K}$ and the "external" energy of a nucleus of

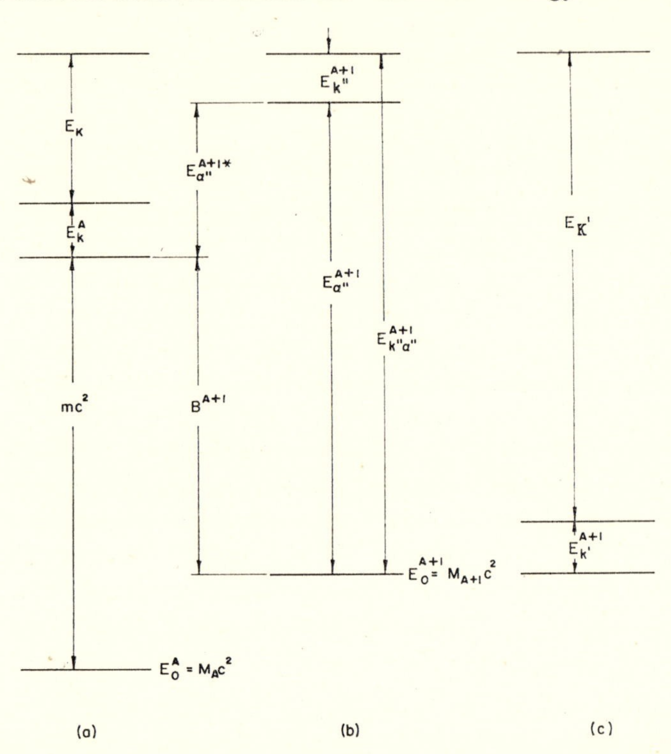

Fig. 3.1. Energy level diagrams for the formation and decay of a compound nucleus; (a) Initial state $|n\rangle$, (b) Intermediate state $|n''\rangle$, (c) Final state $|n'\rangle$.

mass A in "external" state k. Unless stated otherwise, the label k denotes the state of translational motions of the nucleus. We have assumed here that the total energy of a nucleus, $E^A_{k\alpha}$, can be expressed as the sum of its "internal" energy E^A_α and E^A_k. The binding energy, $(m + M_A - M_{A+1})\, c^2$, is represented by B^{A+1}, where all reference to nuclear masses is to ground state rest masses. The various energies that will enter into the present discussion are illustrated in the energy diagram given in Fig. 3.1. In Eq. 3.44 we have extracted from the off-diagonal matrix elements their dependence upon occupation number in much the same manner as in Section D of Chapter II. The relevant part of the potential here is seen to be V^N since the terms that contribute are those describing the emission of a neutron. To illustrate how the sum over neutron occupation numbers is performed, we explicitly display the distribution of the particles among cells in a given m state (see Eqs. 3.35 and 3.36) as

$$\begin{aligned} |m\rangle &= |N'_{\mathbf{X}\varkappa\lambda}\rangle\, |N'_{A\alpha k}\rangle\, |N'(\mathbf{X}_1, \mathbf{K}_1, s_1) \ldots N'(\mathbf{X}_j, \mathbf{K}_j, s_j) \ldots\rangle \\ &= |N''_{\mathbf{X}\varkappa\lambda}\rangle\, |N''_{A\alpha k}\rangle\, |N''(\mathbf{X}_1, \mathbf{K}_1, s_1) \ldots 1 - N''(\mathbf{X}_j, \mathbf{K}_j, s_j) \ldots\rangle \end{aligned} \tag{3.44a}$$

where we have written the photon and neutron occupation numbers relative to the intermediate state, $|n''\rangle = |N''_{\mathbf{X}\varkappa\lambda}\rangle\, |N''_{A\alpha k}\rangle\, |N''_{\mathbf{X}\mathbf{K}s}\rangle$. The photon distribution is not changed since we are considering only V^N, and in this particular case the neutron with spin s_j at the phase point $(\mathbf{X}_j, \hbar\mathbf{K}_j)$ is being emitted. So far as the sum over neutron occupation number is concerned the m sum now becomes a sum over $\mathbf{X}_j$, $\mathbf{K}_j$ and s_j, since any of the neutrons present in any m state can be emitted. We can immediately set $\mathbf{X}_j = \mathbf{X}$ because only the neutrons at $\mathbf{X}$ are of interest and emission is presumed to take place at the point of interaction. Thus the dependence upon neutron occupation number for the emission process is simply $[1 - N(\mathbf{X}_j, \mathbf{K}_j, s_j)]$. We have also attempted to show explicitly in Eq. 3.44 those degrees of freedom which will influence the matrix elements of V^N. Since in the state $|m\rangle$ the nucleus is in its "internal" ground state, the dependence upon α_j is suppressed.

In a similar manner we may display the contributions from states like the final state as

$$\sum_{k_j\varkappa_j\lambda_j} \frac{|V^{\gamma}_{k''\alpha'',k_j\varkappa_j\lambda_j}|^2\,[1 + N(\mathbf{X}_j, \boldsymbol{\kappa}_j, \lambda_j)]}{E^A_k + E_K + B^{A+1} - E^{A+1}_{k_j} - E_{\kappa_j}} \tag{3.45}$$

where $E_\varkappa$ represents the energy of a photon with momentum $\hbar\boldsymbol{\kappa}$, and the contributions from states like the intermediate state as

$$\sum_{\substack{k_j\alpha_j\mathbf{K}'_j\mathbf{K}'_l s'_j s'_l \\ \neq(k''\alpha''\mathbf{K}''_j\mathbf{K}''_l s''_j s''_l)}} \frac{|V^N_{k''\alpha'',k_j\alpha_j\mathbf{K}'_j s'_j\mathbf{K}'_l s'_i}|^2\, N(\mathbf{X}, \mathbf{K}'_j, s'_j)\,[1 - N(\mathbf{X}, K'_l, s'_l)]}{E^A_k + E_K + B^{A+1} - E^{A+1}_{\alpha_j} - E^{A+1}_{k_j} - E_{K'_l} + E_{K_j}} \tag{3.46}$$

On account of their denominators, the terms in Eq. 3.46 make a negligible contribution. They describe an increase in the width and shift function due to a scattering interaction (elastic if $E_{\alpha_j}^{A+1}$ is equal to $E_{\alpha''}^{A+1}$) between the compound nucleus and the neutron field. We expect the effects of this type of collision broadening to be relatively small, and will ignore such terms in the following. The contributions to the m sum in Eq. 3.43, to a good approximation, are then given by expressions in Eqs. 3.44 and 3.45.

It will be convenient to replace the momentum sums by appropriate integrals. For typical quantization cells with characteristic length $L \approx 10^{-4}$ cm, the momentum uncertainty $\Delta K/K$ (or $\Delta\varkappa/\varkappa$) is of order 10^{-3} for 10^{-3} eV neutrons (or keV photons). Thus in all practical cases the distribution of points in momentum space may be treated as essentially continuous. The second term in Eq. 3.43 then becomes

$$\sum_{k_j s_j} \left(\frac{L}{2\pi}\right)^3 \int \mathrm{d}^3 K_j \frac{|V^N_{k''\alpha'',k_j\mathbf{K}_j s_j}|^2 [1 - N(\mathbf{X}, \mathbf{K}_j, s)]}{E_K + E_k^A - E_{K_j} - E_{k_j}^A}$$

$$+ \sum_{k_j\lambda_j} \left(\frac{L}{2\pi}\right)^3 \int \mathrm{d}^3\varkappa \frac{|V^\gamma_{k''\alpha'',k_j\varkappa\lambda_j}|^2 [1 + N(\mathbf{X}, \boldsymbol{\kappa}, \lambda)]}{E_k^A + E_K + B^{A+1} - E_{k_j}^{A+1} - E_\kappa} \tag{3.47}$$

or

$$\frac{1}{2}\left(\frac{L}{2\pi}\frac{\sqrt{2m}}{\hbar}\right)^3 \sum_{k_j s_j} \int \mathrm{d}\Omega_{K_j} \int_0^\infty \mathrm{d}E_{K_j} E_{K_j}^{1/2} \frac{|V^N_{k''\alpha'',k_j\mathbf{K}_j s_j}|^2 [1 - N(\mathbf{X}, \mathbf{K}_j, s_j)]}{E_K + E_k^A - E_{K_j} - E_{k_j}^A}$$

$$+ \left(\frac{L}{2\pi\hbar c}\right)^3 \sum_{k_j\lambda_j} \int \mathrm{d}\Omega_\varkappa \int_0^\infty \mathrm{d}E_\varkappa E_\varkappa^2 \times$$

$$\times \frac{|V^\gamma_{k''\alpha'',k_j\varkappa\lambda_j}|^2 [1 + N(\mathbf{X}, \boldsymbol{\kappa}, \lambda_j)]}{E_k^A + E_K + B^{A+1} - E_{k_j}^{A+1} - E_\varkappa}$$

The integrals in Eq. 3.47 may appear to be singular; however, we recall from Section A that $\gamma_{n''}(-i\varepsilon_n)$ is to be evaluated in the sense of a limit, i.e.

$$\lim_{x\to 0+} \gamma_{n''}(x - i\varepsilon_n)$$

By applying Eq. 3.31 we find that Eq. 3.43 may be written as

$$\frac{\hbar}{2}\gamma_{k''\alpha''} = s_{k''\alpha''} - \frac{i}{2}\Gamma_{k''\alpha''} \tag{3.48}$$

where

$$s_{k''\alpha''} = V_{k''\alpha'',k''\alpha''}$$
$$+ \frac{1}{2}\left(\frac{L}{2\pi}\frac{\sqrt{2m}}{\hbar}\right)^3 \sum_{k_j s_j} \int \mathrm{d}\Omega_{K_j} \mathscr{P} \int_0^\infty \mathrm{d}E_{K_j} E_{K_j}^{1/2} \frac{|V^N_{k''\alpha'',k_j\mathbf{K}_j s_j}|^2 [1 - N(\mathbf{X}, \mathbf{K}_j, s_j)]}{E_K + E_k^A - E_{K_j} - E_{k_j}^A}$$
$$+ \left(\frac{L}{2\pi\hbar c}\right)^3 \sum_{k_j\lambda_j} \int \mathrm{d}\Omega_\varkappa \mathscr{P} \int_0^\infty \mathrm{d}E_\varkappa E_\varkappa^2 \frac{|V^\gamma_{k''\alpha'',k_j\varkappa\lambda_j}|^2 [1 + N(\mathbf{X}, \boldsymbol{\kappa}, \lambda_j)]}{E_k^A + E_K + B^{A+1} - E_{k_j}^{A+1} - E_\varkappa} \tag{3.49}$$

$$\tfrac{1}{2}\Gamma_{k''\alpha''}$$
$$= \frac{\pi}{2}\left(\frac{L}{2\pi}\frac{\sqrt{2m}}{\hbar}\right)^3 \sum_{k_j s_j} \int \mathrm{d}\Omega_{K_j} \{|V_{k''\alpha'',k_j\mathbf{K}_j s_j}|^2 \times$$
$$\times [1 - N(\mathbf{X}, \mathbf{K}_j s_j)] E_{K_j}^{1/2}\}_{E_{K_j} = E_K + E_k^A - E_{kj}^A} + \pi\left(\frac{L}{2\pi\hbar c}\right)^3 \times$$
$$\times \sum_{k_j\lambda_j} \int \mathrm{d}\Omega_\varkappa \{|V^\gamma_{k''\alpha'',k_j\varkappa\lambda_j}|^2 [1 + N(\mathbf{X}, \boldsymbol{\kappa}, \lambda_j)] E_\varkappa^2\}_{E_\varkappa = E_k^A + E_K + B^{A+1} - E_{kj}^{A+1}} \tag{3.50}$$

The function γ is now expressed in terms of its real and imaginary parts. When entered into Eq. 3.42, $s_{k''\alpha''}$ gives rise to a displacement of the resonance line the width of which is determined by $\Gamma_{k''\alpha''}$. In our discussion we shall merely note the existence of $s_{k''\alpha''}$ and will not be concerned with its effects. On the other hand, the existence of $\Gamma_{k''\alpha''}$ is obviously crucial in the development of a theory of resonance reactions, and we will shortly return to more discussion of this quantity.

Entering Eq. 3.48 into Eq. 3.42 and again extracting neutron and photon number dependence from the matrix elements, we can exhibit the capture reaction matrix as

$$R^c_{k'\varkappa'\lambda',k\mathbf{K}s} = N(\mathbf{X}, \mathbf{K}, s)\,[1 + N(\mathbf{X}, \boldsymbol{\kappa}', \lambda')]\, r^c_{k'\varkappa'\lambda',k\mathbf{K}s} \tag{3.51}$$

$$r^c_{k'\varkappa'\lambda',k\mathbf{K}s} = \frac{2\pi}{\hbar}\left|\sum_{k''\alpha''} \frac{V^\gamma_{k'\varkappa'\lambda',k''\alpha''}\, V^N_{k''\alpha'',k\mathbf{K}s}}{E_{\alpha''}^{A+1*} + s_{k''\alpha''} + E_{k''}^{A+1} - E_K - E_k^A - \dfrac{i}{2}\Gamma_{k''\alpha''}}\right|^2 \tag{3.52}$$

The energy difference $E_{\alpha''}^{A+1} - B^{A+1}$ is denoted here as $E_{\alpha''}^{A+1*}$. Note that we are labeling the matrix elements again by state labels rather than by occupation numbers. This is because the sum over occupation numbers in the final state in Eq. 3.39 actually reduces to a sum over states as in the earlier cases. The sum over occupation numbers in the initial state is to be carried out formally according to the specific dependence indicated in Eq. 3.51.

Thus far we have not given any specific consideration to the matrix elements of V^{γ} and V^{N}. The discussion of V^{γ} can be made more quantitative and will be considered first. The portion of the interaction between nuclei and electromagnetic field which describes single photon emission or absorption is [11]

$$V^{\gamma} = \sum_{l,L} \frac{e_L}{m_L c} \mathbf{A}(\mathbf{r}_l^L) \cdot \mathbf{p}_l^L \tag{3.53}$$

where the l specifies the nucleus and L specifies the nucleon. The momentum of the nucleon is denoted by $\mathbf{p}$ and the vector field $\mathbf{A}$ represents the transverse radiation field quantized in a manner wholly similar to the quantization of the neutron field in Chapter II. The matrix elements of V^{γ} then become

$$\begin{aligned} V^{\gamma}_{k'\kappa'\lambda',k''\alpha''} &= \left\langle k' \, 0 \left| \sum_{l,L} \frac{e_L}{m_L c} e^{-i\kappa'\cdot \mathbf{r}_l^L} \mathbf{a}_{\lambda'}(\boldsymbol{\kappa}') \cdot \mathbf{p}_l^L \right| k''\alpha'' \right\rangle \\ &\approx \sum_l \langle k' | \, e^{-i\kappa'\cdot \mathbf{R}_l} \, | k'' \rangle \, U^{Rl}_{0\alpha''}(\boldsymbol{\kappa}'\lambda') \end{aligned} \tag{3.54}$$

where

$$U^{Rl}_{0\alpha''}(\boldsymbol{\kappa}\lambda) = \left\langle 0 \left| \sum_L \frac{e_L}{m_L c} e^{-i\kappa\cdot \boldsymbol{\varrho}_l^L} \mathbf{a}_{\lambda}(\boldsymbol{\kappa}) \cdot \mathbf{p}^L \right| \alpha'' \right\rangle \tag{3.55}$$

The label 0 is used to denote the ground "internal" state of the nucleus. To arrive at the above factorization we have introduced the center-of-mass position vector, $\mathbf{R}$, and the relative displacement, $\boldsymbol{\rho}$, so that $\mathbf{r}_l^L = \mathbf{R}_l + \boldsymbol{\rho}_l^L$. These coordinates, however, are not independent. The momentum $\mathbf{p}_l^L$ is conjugate to $\mathbf{r}_l^L$, and therefore consists of contributions from center-of-mass motions as well as from relative motions. But because the nuclear momentum is very small compared to the nucleonic momentum we have neglected the former and set $\mathbf{p}_l^L \approx \mathbf{p}^L$. It is only in this approximate sense that we may isolate the effects due to external medium. The factor $U^{Rl}_{0\alpha''}$ now depends solely upon internal motions and describes the response of the nucleons to the photon field.

Since a particular Fourier component of the neutron field is involved in the matrix element, $V^{N}_{k''\alpha'',k\mathbf{K}s}$, and since the range of neutron-nuclear forces is small compared to the dimension of the quantization cell, it is expected that the matrix elements describing neutron absorption decompose in a fashion similar to the factorization of the photon-emission matrix elements in Eq. 3.54. We shall therefore write

$$V^{N}_{k''\alpha'',kKs} \approx \sum_{l} \langle k'' | e^{iK\cdot R_l} | k \rangle \, U^{Nl}_{\alpha''0}(Ks) \tag{3.56}$$

Both matrix elements of V^{γ} and V^{N} are seen to contain the sum over nuclei. These sums, however, will not appear in the calculation of the reaction matrix (3.52). This is because such a reaction matrix is intended to describe the evolution of the system from a state characterized by a certain number of neutrons, photons, nuclei of masses A and $(A + 1)$ to a state characterized by one less neutron, one more photon, one less mass A nucleus, and one more mass $(A + 1)$ nucleus. The nucleus which absorbs the neutron must be the same nucleus as that which emits the photon, thus elements of the reaction matrix between specified initial and final states will depend only upon the properties of a single nucleus.

The reduced reaction matrix for capture now becomes a sum of matrices each appropriate to an individual nucleus. For the nucleus designated by the label l we have

$$r^{cl}_{k'\varkappa'\lambda',kKs} = \frac{2\pi}{\hbar} \left| \sum_{k''\alpha''} \frac{U^{Rl}_{0\alpha''}(\varkappa'\lambda') U^{Nl}_{\alpha''0}(Ks) \langle k' | e^{-i\varkappa'\cdot R_l} | k'' \rangle \langle k'' | e^{iK\cdot R_l} | k \rangle}{\mathscr{E}_{\alpha''} + E^{A+1}_{k''} - E_K - E^{A}_{k} - \dfrac{i}{2} \Gamma_{\alpha''}} \right|^2 \tag{3.57}$$

where we have ignored the dependence of the level width and level shift upon the external degrees of freedom of the nucleus, and where

$$\mathscr{E}_{\alpha''} = E^{A+1}_{\alpha''} - B^{A+1} + s_{\alpha''} \tag{3.58}$$

is the energy of the αth level in the nucleus of mass $(A + 1)$ as seen by a free neutron in the laboratory. If we assume for illustrative purposes that the nuclei in the system are characterized by well-separated energy levels,* then Eq. 3.57 reduces to a sum of a single-level resonances

$$r^{cl}_{k'\varkappa'\lambda',kKs} \approx \frac{2\pi}{\hbar} \sum_{\alpha''} |U^{Rl}_{0\alpha''}(\varkappa'\lambda') \, U^{Nl}_{\alpha''0}(Ks)|^2 \times \\ \times \left| \sum_{k''} \frac{\langle k' | e^{-i\varkappa'\cdot R_l} | k'' \rangle \langle k'' | e^{iK\cdot R_l} | k \rangle}{\mathscr{E}_{\alpha''} + E^{A+1}_{k''} - E_K - E^{A}_{k} - \dfrac{i}{2} \Gamma_{\alpha''}} \right|^2 \tag{3.59}$$

* In the conventional theory of resonance[1,6] one introduces a level-spacing D which represents the average separation between neighboring resonance levels. Values of D range from several hundred KeV for light elements down to a few eV for $A \approx 100$, and will in general decrease with increasing excitation energy. Thus it is meaningful to speak of isolated resonance levels only if $\Gamma_{\alpha''} \ll D$.

The matrix elements $U^{R}_{0\alpha''}$ and $U^{N}_{\alpha''0}$ incorporate all the responses in the interior of the nucleus to the reaction, and are complicated quantities which cannot be discussed quantitatively in the present development. For our purposes it is sufficient to replace them by more familiar quantities. We observe that the level width given in Eq. 3.50 can be identified as a sum of partial widths appropriate to the decay of compound nucleus by either neutron or photon emission. Specifically the radiation width for the α''th level is

$$\frac{1}{2}\Gamma^{(R)}_{\alpha''} \approx \pi\left(\frac{L}{2\pi\hbar c}\right)^3 \sum_{k_j\lambda_j} \int d\Omega_\varkappa \{|\langle k''|\, e^{-i\varkappa\cdot\mathbf{R}_l}\, |k_j\rangle|^2 \times$$

$$\times\, |U^{Rl}_{\alpha''0}(\varkappa\lambda_j)|^2\, [1 + N(\mathbf{X}, \varkappa, \lambda_j)]\, E^2_\varkappa\}_{E_\varkappa = E^A_k + E_K + B^{A+1} - E^{A+1}_{k_j}} \tag{3.60}$$

where use has been made of Eq. 3.54. We will assume that we may ignore the factor $[1 + N(\mathbf{X}, \kappa, \lambda_j)]$. If we further assume that the difference in "external" energies, $E^A_k - E^{A+1}_{kj}$, is negligible compared to the excitation energy of the compound nucleus, then the sum over k_j may be performed to give

$$\frac{1}{2}\Gamma^{(R)}_{\alpha''} \approx \pi\left(\frac{L}{2\pi\hbar c}\right)^3 (E_K + B^{A+1})^2 \sum_{\lambda} \int d\Omega_\varkappa\, |U^{Rl}_{\alpha''0}(\varkappa\lambda)|^2 \tag{3.61}$$

Using similar arguments and approximations we find the first term in Eq. 3.50 to be given by

$$\frac{1}{2}\Gamma^{(N)}_{\alpha''} \approx \frac{\pi}{2}\left(\frac{L}{2\pi}\frac{\sqrt{2m}}{\hbar}\right)^3 \sqrt{E_K} \sum_{s} \int d\Omega_{K'}\, |U^{Nl}_{\alpha''0}(\mathbf{K}'_s)|^2 \tag{3.62}$$

which can be identified as the neutron width. In the sense of the above approximations and if $|U^{R_l}|^2$ and $|U^{N_l}|^2$ can be considered as constants these results show that the radiation width is essentially energy independent, whereas the neutron width is proportional to the neutron speed.* Eqs. 3.61 and 3.62 are useful in that they allow us to write the

* For the case of U^{235} see Oleksa.[16] Because of its dependence upon $(B^{A+1})^2$ the radiation width can be expected to decrease as A increases. The energy dependence of the neutron width is in agreement with the conventional results[1] for neutrons of zero angular momentum and therefore implies that $|U^{Nl}|^2$ can indeed be treated as a constant so long as the neutron energy is not so high that neutrons with higher angular momentum begin to interact appreciably.

elements of the reaction matrix in terms of level widths, and in the present treatment the latter quantities will be treated as empirical parameters. It is expected that $|U^{Rl}_{\alpha''0}|^2$ is quite insensitive to the directions of **K**, so that we have*

$$\frac{1}{2}\Gamma^{(N)}_{\alpha''} \approx 2\pi^2\left(\frac{L}{2\pi}\frac{\sqrt{2m}}{\hbar}\right)^3 \sqrt{E_K}\sum_s |U^{Nl}_{\alpha''0}(\mathbf{K}s)|^2 \tag{3.63}$$

The same may be said for the dependence of $|U^{Rl}_{\alpha''0}|^2$, although, as we will show later, the assertion is not necessary in this case.

Further progress from this point, at least so far as the reduction of Eq. 3.59 to useful forms is concerned, requires specific assumptions regarding the macroscopic state of the system. It will be necessary to know whether the external degrees of freedom of the nuclei are those appropriate to a system in solid, liquid, or gaseous state in order to compute the indicated matrix elements. These matters will be considered in the following chapter. In concluding this section we shall examine some of the more general aspects of the collision terms in the balance relation which describe the effect of radiative capture processes. These terms now appear in Eq. 3.39 as

$$V^{-1}\sum_{nn's} W^{c}_{n'n} D_{nn}(t) = V^{-1}\sum_{lk'\varkappa'\lambda'ns} N(\mathbf{X}, \mathbf{K}, s)\,[1 + N(\mathbf{X}, \boldsymbol{\varkappa}', \lambda')] \times$$
$$\times\, r^{cl}_{k'\varkappa'\lambda'k\mathbf{K}s} D_{nn}(t)\,\delta(E^{A+1}_{k'} - B^{A+1} + E_{\varkappa'} - E^{A}_{k} - E_K) \tag{3.64}$$

Evidently the n sum leads to functionals of various doublet densities. However, to avoid explicit consideration of these higher-order densities, we shall liberally (and for the moment uncritically) replace averages of functions by functions of averages.† Thus,

$$V^{-1}\sum_{nn's} W^{c}_{n'n} D_{nn}(t) \approx \sum_{lk'\varkappa'\lambda's} F_s(\mathbf{X}, \mathbf{K}, t)\,[1 + F_{\lambda'}(\mathbf{X}, \varkappa', t)] \times$$
$$\times\, r^{cl}_{k'\varkappa'\lambda'k\mathbf{K}s} D_{kk}(t)\,\delta(E^{A+1}_{k'} - B^{A+1} + E_{\varkappa'} - E^{A}_{k} - E_K) \tag{3.65}$$

---

* This is equivalent to the assumption that neutron emission or absorption is essentially spherically symmetric, a condition usually valid at least for $E_K \lesssim 100$ KeV.[6]

† Had we retained the doublet densities then Eq. 3.39, which may be regarded as an equation for the singlet density, would be incomplete for the determination of $F(\mathbf{X}, \mathbf{K}, t)$. An equation for the doublet density is therefore necessary, and we will find that it contains the triplet densities. Hence, an infinite set of coupled equations is generated.

where $F_s(\mathbf{X}, \mathbf{K}, t)$ is the expected number of neutrons per unit volume at time t with spin s and momentum $\hbar\mathbf{K}$ at $\mathbf{X}$, $F_\lambda(\mathbf{X}, \boldsymbol{\kappa}, t)$ is the expected number of photons at time t with polarization λ and momentum $\hbar\boldsymbol{\kappa}$ at $\mathbf{X}$, and $D_{kk}(t)$ is the probability of finding the target nucleus in the state k at time t. For most applications involving the neutron transport equation the neutron spin orientation is not a variable of interest,* so that there will be no loss of generality if we assume the spins are randomly distributed, or

$$F_s(\mathbf{X}, \mathbf{K}, t) = \tfrac{1}{2} F(\mathbf{X}, \mathbf{K}, t) \tag{3.66}$$

Now Eq. 3.65 becomes

$$V^{-1} \sum_{nn's} W^c_{n'n} D_{nn}(t) \approx F(\mathbf{X}, \mathbf{K}, t) \sum_{lk'\varkappa'\lambda's} [1 + F_{\lambda'}(\mathbf{X}, \varkappa', t)] \times$$
$$\times \tfrac{1}{2} r^{cl}_{k'\varkappa'\lambda', k\mathbf{K}s}\, \delta(E^{A+1}_{k'} + E_{\varkappa'} - B^{A+1} - E^A_k - E_K)\, D_{kk}(t) \tag{3.67}$$

The capture contribution is thus in a conventional form of a reaction rate times the neutron density. In the following chapter we shall show how this reaction rate can be reduced to the more familiar expressions for the cross section.

C. Elastic Scattering

For neutrons with energies below the inelastic scattering threshold, about 1 MeV for light nuclei down to $\approx$100 keV for high A, the only process available for their energy moderation is elastic scattering.† The neutron energy distribution as determined from the transport equation can be quite sensitive to the energy-transfer mechanisms underlying this type of collision. The fact that the neutron scattering can be significantly influenced by the atomic motions of the system not only introduces additional complexities into the transport equation at low energies, but also suggests the use of neutrons as an effective probe for the study of solids and liquids. These remarks will be elaborated in

---

* A possible exception could be the case of neutron transport in inhomogeneous magnetic field. Admittedly this is not a system of practical interest.

† For a discussion of the slowing down of neutrons by elastic collisions see Marshak[17] and Ferziger and Zweifel.[18]

greater detail in the next chapter on the basis of the development presented in this section.

There are two types of elastic scattering processes which should be distinguished at the outset since they will require somewhat different treatments. The first process is like radiative capture in that a compound nucleus is formed, but rather than decaying by the emission of a photon the compound nucleus decays to ground state by the emission of a neutron. This reaction is known as elastic resonant scattering. The second process is a direct reaction known as potential scattering, which can be considered as taking place in the immediate vicinity of the surface of the nucleus so that there is effectively no penetration.* In general, potential scattering dominates in energy regions away from any resonance, whereas within the vicinity of the resonance peak resonant scattering dominates. In regions where both kinds of scattering are of the same strength it is known that appreciable interference can exist, which is generally destructive at the low-energy side and constructive at the high-energy side.† We shall therefore consider both processes at the same time in order to include such interference effects in the present analysis.

The reaction matrix describing the scattering interaction is again given by Eq. 3.34 where now only V^N, the nuclear part of the potential, needs to be considered. Here the class of initial and final states is that characterized by the conservation of neutrons, photons, and nuclei. There are, however, two sub-classes corresponding to the increase and decrease respectively of a neutron at the phase point of interest. In the binary collision they constitute the scattering gain and loss to the balance relation as indicated in the qualitative discussion given in Chapter II.

For the treatment of both direct and resonance processes we assume that V^N has nonvanishing matrix elements between initial and final states as well as between intermediate state and final or initial state. The reaction matrix can be written in a form similar to Eq. 3.51 and 3.52,

$$R^{Es}_{k'\mathbf{K}'s',\,k\mathbf{K}s} = N(\mathbf{X}, \mathbf{K}, s)\,[1 - N(\mathbf{X}, \mathbf{K}', s')]\, r^{Es}_{k'\mathbf{K}'s',\,k\mathbf{K}s} \qquad (3.68)$$

---

* Cf. reference 1, p. 393; see also remarks by Lane and Thomas, reference 2, p. 261.

† A rather striking example of this phenomenon is the sulfur resonance line at ≈100 KeV (also the silicon line at ≈150 KeV).[19]

$$r^{Es}_{k'\mathbf{K}'s',k\mathbf{K}s} = \frac{2\pi}{\hbar}\left| V^{N}_{k'\mathbf{K}'s',k\mathbf{K}s} - \sum_{\alpha''k''} \frac{V^{N}_{k'\mathbf{K}'s',k''\alpha''}\, V^{N}_{k''\alpha'',k\mathbf{K}s}}{E^{A+1*}_{\alpha''} + s_{k''\alpha''} + E^{A+1}_{k''} - E^{A}_{k} - E_{K} - \frac{i}{2}\Gamma_{k''\alpha''}} \right|^2 \tag{3.69}$$

These two expressions are appropriate to collisions resulting in "scattering loss". Corresponding expressions for "scattering gain" are obtained by merely interchanging the state labels (k, $\mathbf{K}$, s) and (k', $\mathbf{K}'$, s'). The various energies appearing in Eq. 3.69 are the same as those introduced in the previous section. The matrix elements of V^N may again be factored as indicated in Eq. 3.56, and we obtain

$$r^{Es}_{k'\mathbf{K}'s',k\mathbf{K}s} = \frac{2\pi}{\hbar}\left| V^{N}_{k'\mathbf{K}'s',k\mathbf{K}s} - \sum_{l\alpha''k''} \frac{U^{Nl}_{0\alpha''}\, U^{Nl}_{\alpha''0} \langle k'|\, \mathrm{e}^{-i\mathbf{K}'\cdot\mathbf{R}_l}\, |k''\rangle \langle k''|\, \mathrm{e}^{i\mathbf{K}\cdot\mathbf{R}_l}\, |k\rangle}{\mathscr{E}_{\alpha''} + E^{A+1}_{k''} - E^{A}_{k} - E_{K} - \frac{i}{2}\Gamma_{\alpha''}} \right|^2 \tag{3.70}$$

where we have introduced $\mathscr{E}_{\alpha''}$ according to Eq. 3.58.

The direct matrix elements, $V^{N}_{k'\mathbf{K}'s',k\mathbf{K}s}$, can be estimated in terms of a specific model of a neutron-nuclear interaction. It is to be emphasized that this use of a model does not affect the other matrix elements of V^N, those describing resonance scattering. This will become evident in the following, for the parameters of the model are to be determined according to comparisons with data from low-energy potential scattering. Since these parameters are fitted to experiments in the sense of certain calculational approximations, i.e., the Born approximation or first-order perturbation theory as presented here, it is not clear that such a model should be employed in general theoretical analyses which are not restricted to these approximations. Conversely, it will be seen that the model employed in the context of first-order perturbation theory can be adjusted to describe the experimental results exceedingly well, and its use enables one to explore the specifically macroscopic medium effects with reasonable confidence.

The model, as we shall construe it, is introduced formally as the potential,

$$V = \sum_l \int d^3x \, [\psi_j^+(\mathbf{x}) \, \psi_j(\mathbf{x}) \, v_0^l(|\mathbf{x} - R_l|) + \mathbf{I}_l \cdot \psi_j^+(\mathbf{x}) \, \sigma_{jk} \, \psi_k(\mathbf{x}) \, v_1^l(|\mathbf{x} - \mathbf{R}_l|)] \tag{3.71}$$

The spinor field operator $\psi_j(\mathbf{x})$ has been discussed in the previous chapter. The components of the vector $\boldsymbol{\sigma}$ are the Pauli spin matrices[11] and consequently $\sigma/2$ represents the intrinsic spin (in units of $\hbar$) of the neutron. The vector $\mathbf{I}_l$ represents the observed angular momentum of the lth nucleus in its ground state. Instead of the delta function (pseudopotential) introduced by Fermi,[20,21]* we shall depart slightly from convention and suggest the use of Yukawa functions to represent the short-range potentials v_i^l, $i = 0, 1$,

$$v_i^l(r) = g_i^l \frac{e^{-r/\lambda_i^l}}{r} \tag{3.72}$$

The reasons for our preference for this potential function will be made more explicit later; for the moment we simply remark that it provides a cross section for neutron-nuclear potential scattering with a somewhat expanded range of qualitative validity.

A more useful expression of V is obtained by using the spinor field expansion in Eq. 2.8. We find

$$V = \sum_{\mathbf{XKK}'ss'} a^+(\mathbf{X}, \mathbf{K}, s) \, a(\mathbf{X}, \mathbf{K}', s') \sum_l U_l(-\mathbf{Q}, s, s') \, e^{-i\mathbf{Q}\cdot\mathbf{R}_l} \tag{3.73}$$

where

$$\mathbf{Q} = \mathbf{K} - \mathbf{K}', \tag{3.74}$$

$$U_l(\mathbf{Q}, s, s') = V^{-1} \int d^3R \, E(\mathbf{X}, \mathbf{R} + \mathbf{R}_l) \, e^{i\mathbf{Q}\cdot\mathbf{R}} \times [v_0^l(R) \, \delta_{ss'} + \mathbf{I}_l \cdot u_j^+(s) \, \boldsymbol{\sigma}_{jk} u_k(s') \, v_1^l(R)] \tag{3.75}$$

The Hermitian character of this coarse-grained potential can be readily demonstrated, for

$$V^+ = \sum_{\mathbf{XKK}'ss'} a^+(\mathbf{X}, \mathbf{K}', s') \, a(\mathbf{X}, \mathbf{K}, s) \sum_l U_l^+(-\mathbf{Q}, s, s') \, e^{i\mathbf{Q}\cdot\mathbf{R}_l} \tag{3.76}$$

and since $\mathbf{S}_l$ and $\boldsymbol{\sigma}$ are Hermitian, and v_i^l is real, we have

$$\begin{aligned} U_l^+(\mathbf{Q}, s', s) &= V^{-1} \int d^3R \, E(\mathbf{X}, \mathbf{R} + \mathbf{R}_l) \, e^{-i\mathbf{Q}\cdot\mathbf{R}_l} \times [v_l^0(R) \, \delta_{ss'} + \mathbf{I}_l \cdot u_j^+(s') \, \boldsymbol{\sigma}_{jk}^* \, u_k(s) \, v_1^l(R)] \\ &= U_l(-\mathbf{Q}, s, s') \end{aligned} \tag{3.77}$$

---

* The Fermi approximation has been recently studied by Plummer[22] and Summerfield.[23]

Thus by interchanging (**K**s) and (**K**$'$, s') in Eq. 3.76 we obtain $V^+ = V$, which is of course a necessary property of the potential.

The inclusion of spin-dependent interactions, as represented by the second term in (3.71) or (3.75), necessitates a slight modification of our treatment of the matrix elements. Thus far, the labels k and α have been used to specify the "external" and "internal" states of a nucleus, and the possible existence of a nuclear spin has not been considered. However, the presence of a spin-dependent term in V^N now requires the introduction of a label to specify the spin states of nuclei in their ground states. This additional complexity can be avoided in many cases where the results, though strictly applicable only to spinless nuclei, are still of sufficiently general interest. Thus except for the section in the next chapter dealing specifically with potential scattering we shall assume that U_l is adequately characterized by only the first term in (3.75).

The direct matrix elements are now of the form

$$V^N_{k'\mathbf{K}'s',k\mathbf{K}s} = \sum_l \langle k'|\, U_l(\mathbf{Q}, s, s')\, e^{i\mathbf{Q}\cdot\mathbf{R}}\, |k\rangle \tag{3.78}$$

and the elements of the reduced transition matrix for elastic scattering, with this particular choice of a potential, are

$$r^{Es}_{k'\mathbf{K}'s',k\mathbf{K}s} = \frac{2\pi}{\hbar}\Bigg|\sum_l \langle k'|\, U_l(\mathbf{Q}, s, s')\, e^{i\mathbf{Q}\cdot\mathbf{R}_l}\, |k\rangle - \sum_{l\alpha''k''} \frac{U^{Nl}_{0\alpha''}(\mathbf{K}'s')\, U^{Nl}_{\alpha''0}(\mathbf{K}s)\, \langle k'|\, e^{-i\mathbf{K}'\cdot\mathbf{R}_l}\, |k''\rangle \langle k''|\, e^{i\mathbf{K}\cdot\mathbf{R}_l}\, |k\rangle}{\mathscr{E}_{\alpha''} + E^{A+1}_{k''} - E^A_k - E_K - \dfrac{i}{2}\Gamma_{\alpha''}}\Bigg|^2 \tag{3.79}$$

Although still rather formal, this expression will provide a suitable starting point for explicit considerations of medium effects and for the introduction of useful assumptions and approximations. We shall close this section by noting that the elastic scattering contributions to the collision terms in Eq. 3.39 may be displayed in a form analogous to Eq. 3.67

$$\begin{aligned} V^{-1}\sum_{nn's} (W^{EsG}_{n'n} - W^{EsL}_{n'n})\, D_{nn}(t) &\approx \left[1 - \frac{V}{2}F(\mathbf{X}, \mathbf{K}, t)\right] \sum_{\mathbf{K}'} F(\mathbf{X}, \mathbf{K}', t) \sum_{k's'ks} \frac{1}{2} r^{Es}_{k\mathbf{K}s,k'\mathbf{K}'_s} \times \\ &\times \delta(E^A_k + E_K - E^A_{k'} - E_{K'})\, D_{k'k'}(t) \\ &- F(\mathbf{X}, \mathbf{K}, t) \sum_{\mathbf{K}'} \left[1 - \frac{V}{2}F(\mathbf{X}, \mathbf{K}', t)\right] \sum_{k's'ks} \frac{1}{2} r^{Es}_{k'\mathbf{K}'s',k\mathbf{K}s} \times \\ &\times \delta(E^A_k + E_K - E^A_{k'} - E_{K'})\, D_{kk}(t) \end{aligned} \tag{3.80}$$

D. Fission and Inelastic Scattering

The contributions to the balance relation (3.39) from fission and inelastic scattering, like those from elastic scattering, also can be expressed as gains and losses, the fission loss term combining with the radiative capture loss to account for the removal of neutrons by absorption. The processes of fission and inelastic scattering are more complicated in that they should be properly regarded as multiple-stage (at least three) reactions, for it is likely in fission that the compound nucleus decays to two excited fragments which then subsequently decay to less excited fragments via neutron and/or photon emission. In inelastic scattering, the compound nucleus emits a neutron, leaving an excited nucleus which will de-excite by photon emission after some time lag.* Thus a reaction theory capable of describing three-stage events is necessary to treat these processes. The reaction matrix given in Eq. 3.34 can be used to describe only first- and second-order processes, so that higher-order terms would have to be included in Eqs. 3.11 and 3.12. The generalization of the formalism in Section A is a straightforward matter; however,

$$R^{Fl}_{k_1\alpha_1,k\mathbf{K}s}(\{Ks\}_J) = N(\mathbf{X},\mathbf{K},s)\prod_{\{Ks\}_J}{}^{*}\,[1 - N(\mathbf{X},\mathbf{K}_j,s_j)]\, r^{Fl}_{k_1\alpha_1,k\mathbf{K}s}(\{Ks\}_J), \tag{3.81}$$

$$r^{Fl}_{k_1\alpha_1,k\mathbf{K}s}(\{Ks\}_J) = \frac{2\pi}{\hbar}\left|\sum_{k''\alpha''}\frac{V^N_{k_1\alpha_1,k''\alpha''}(\{Ks\}_J)\,V^N_{k''\alpha'',k\mathbf{K}s}}{\mathscr{E}_{\alpha''} + E^{A+1}_{k''} - E^A_k - E_K - \dfrac{i}{2}\Gamma_{\alpha''}}\right|^2 \tag{3.82}$$

In Eq. 3.81 the product is such that whenever $(\mathbf{K}_j, s_j) = (\mathbf{K}, s)$ the factor $[1 - N(\mathbf{X},\mathbf{K},s)]$ is to be replaced by unity. This then ensures that a neutron can be emitted having the same momentum and spin as that of the neutron initially absorbed. The matrix element describes the formation of a compound nucleus and has been encountered previously. Its dependence upon "external" degrees of freedom is given by Eq. 3.56. The other matrix element, $V^N_{k_1\alpha_1,k''\alpha''}(\{Ks\}_J)$, represents the "falsification" mentioned above since it describes the decay of the compound

---

* Inelastic scatterings in general need not involve the formation of a compound nucleus. A number of direct reactions of this type have been cited in Lane and Thomas, reference 2, p. 264. The direct scattering by a rotating nonspherical nucleus has been investigated by Chase, *et al.*[24]

nucleus into two fission fragments in external states k_1 and internal states α_1, and J neutrons with momentum and spin distribution $\{Ks\}_J$. In this case an expression similar to Eq. 3.56 for the "external state" dependence is not to be expected since neither of the fission fragments may be regarded as located at the centre-of-mass position of the compound nucleus and one has no knowledge of which of J neutrons is emitted by a given fragment.

The fission contributions to the balance relation can be decomposed into gain and loss terms depending upon whether or not the distribution $\{Ks\}_J$ contains the momentum and spin of interest. In the sense of approximations inherent in Eq. 3.80, they may be displayed as

$$V^{-1} \sum_{nn's} [N'(\mathbf{X}, \mathbf{K}, s) - N(\mathbf{X}, \mathbf{K}, s)] \, W^F_{n'n} D_{nn}(t)$$

$$\approx \sum_{\substack{lJk_1\alpha_1 \\ \mathbf{K}'k's'}} \sum_{\{\mathbf{K}\}JK} F(\mathbf{X}, \mathbf{K}', t) \prod_{\{\mathbf{K}\}JK}{}^{*} \left[1 - \frac{V}{2} F(\mathbf{X}, \mathbf{K}_j, t)\right] \frac{1}{2} r^{Fl}_{k_1\alpha_1, k'\mathbf{K}'s'} (\{K\}_{JK}) \times$$

$$\times \, D_{k'k'}(t) \, \delta(E^A_{k'} + E_{K'} - E_f)$$

$$- F(\mathbf{X}, \mathbf{K}, t) \sum_{\substack{lJK_1\alpha_1 \\ ks}} \sum_{\{\mathbf{K}\}'_{JK}} \prod_{\{\mathbf{K}\}'_{JK}}{}^{*} \left[1 - \frac{V}{2} F(\mathbf{X}, \mathbf{K}_j, t)\right] \frac{1}{2} r^{Fl}_{k_1\alpha_1, k\mathbf{K}s} (\{K\}'_{JK})$$

$$\times \, D_{kk}(t) \, \delta(E^A_k + E_K - E_f) \tag{3.83}$$

where

$$r^{Fl}_{n'n}(\{K\}) = \sum_{\{s\}_J} r^{Fl}_{n'n}(\{Ks\}_J) \tag{3.84}$$

The distributions $\{K\}_{JK}$ and $\{K\}'_{JK}$ denote those which contain and do not contain the momentum $\mathbf{K}$ respectively. The energy E_f may be obtained as follows. The total energy before fission is

$$\varepsilon_n = (m + M_A)\, c^2 + E^A_k + E_K \tag{3.85}$$

and, according to the present approach, the total energy after fission is

$$\varepsilon_{n'} = (M_1 + Jm)\, c^2 + E_{\alpha_1} + E_{k_1} + E_{\{K\}_J} \tag{3.86}$$

where M_1 is the sum of the rest masses of the two fission fragments, E_{α_1} is their internal energy (subsequently leads to emission of γ, β, and neutrons), E_{k_1} is their "external" (kinetic) energy and $E_{\{K\}_J}$ is the kinetic energy of the fission neutrons. Thus

$$E_f = \varepsilon_{n'} - (m + M_A)\, c^2$$

The sum of the last three terms in Eq. 3.86 gives the energy released by the fission process. For U^{235} this[6] is about 200 MeV, roughly the same as the energy due to the mass difference. Hence from the standpoint of energy conservation there is effectively no fission threshold.* Such will not be the case for inelastic scattering.

We now consider the decay of the compound nucleus by neutron emission; the residual nucleus, being in an excited state, then decays to ground state by gamma emission. To treat this process with the formalism employed throughout this chapter, it will be necessary to assume that a potential exists that has nonvanishing matrix elements between states in which both neutron and photon numbers are changed. We shall denote this potential as V' since it does not conform to the separation (3.41) into parts purely electromagnetic or nuclear. The relevant elements of the transition matrix for inelastic scattering, in the present approximation, are therefore

$$R^{Is}_{k'\mathbf{K}'s'\varkappa'\lambda',k\mathbf{K}s} = N(\mathbf{X},\mathbf{K},s)\,[1 - N(\mathbf{X},\mathbf{K}',s')]\,[1 + N(\mathbf{X},\boldsymbol{\kappa}',\lambda')]\,r^{Is}_{k'\mathbf{K}'s'\varkappa'\lambda',k\mathbf{K}s} \tag{3.87}$$

$$r^{Is}_{k'\mathbf{K}'s'\varkappa'\lambda',k\mathbf{K}s} = \frac{2\pi}{\hbar}\left|\sum_{k''\alpha''}\frac{V'_{k'\mathbf{K}'s'\varkappa'\lambda',k''\alpha''}\,V^N_{k''\alpha'',k\mathbf{K}s}}{\mathscr{E}_{\alpha''} + E^{A+1}_{k''} - E^A_k - E_K - \dfrac{i}{2}\Gamma_{\alpha''}}\right|^2 \tag{3.88}$$

In this case we may expect the matrix elements describing the decay of the compound nucleus to be approximately factorable as in Eqs. 3.54 and 3.56,

$$V_{k'\mathbf{K}'s'\varkappa'\lambda',k''\alpha''} \approx \sum_l \langle k' \mid e^{-i\mathbf{R}_l\cdot(\mathbf{K}'+\varkappa')} \mid k''\rangle\, U^l_{0\alpha''}(\mathbf{K}'s'\boldsymbol{\kappa}'\lambda') \tag{3.89}$$

Since the nucleus emitting the photon and neutron has to be the one that captures the neutron, we can again introduce the reduced reaction matrix appropriate to a single nucleus,

$$r^{Isl}_{k'\mathbf{K}'s'\varkappa'\lambda',k\mathbf{K}s} \approx \frac{2\pi}{\hbar}\left|\sum_{k''\alpha''}\frac{U^l_{0\alpha''}\,U^{Nl}_{\alpha''0}\,\langle k' \mid e^{-i\mathbf{R}_l\cdot(\mathbf{K}'+\varkappa')} \mid k''\rangle\,\langle k''\mid e^{i\mathbf{K}\cdot\mathbf{R}_l} \mid k\rangle}{\mathscr{E}_{\alpha''} + E^{A+1}_{k''} - E^A_k - E_K - \dfrac{i}{2}\Gamma_{\alpha''}}\right|^2 \tag{3.90}$$

---

* See, however, Weinberg and Wigner, reference 6, p. 108, for cases in which A is even.

The contributions to the balance relation due to inelastic scattering now become

$$V^{-1} \sum_{nn's} (W_{n'n}^{IsG} - W_{n'n}^{IsL}) D_{nn}(t)$$

$$\approx \left[1 - \frac{V}{2} F(\mathbf{X}, \mathbf{K}, t)\right] \sum_{lk'\mathbf{K}'s'\varkappa\lambda ks} F(\mathbf{X}, \mathbf{K}', t) [1 + F_\lambda(\mathbf{X}, \boldsymbol{\kappa}, t)] \times$$

$$\times \tfrac{1}{2} r_{k\mathbf{K}s\varkappa\lambda, k'\mathbf{K}'s'}^{IsL} \delta(E_{k'}^A + E_{K'} - E_k^A - E_K - E_\varkappa) D_{k'k'}(t)$$

$$- F(\mathbf{X}, \mathbf{K}, t) \sum_{lk'\mathbf{K}'s'\varkappa'\lambda'ks} \left[1 - \frac{V}{2} F(\mathbf{X}, \mathbf{K}', t)\right] [1 + F_{\lambda'}(\mathbf{X}, \boldsymbol{\kappa}', t)] \times$$

$$\times \tfrac{1}{2} r_{k'\mathbf{K}'s'\varkappa'\lambda', k\mathbf{K}s}^{Isl} \delta(E_{k'}^A + E_{K'} + E_{\varkappa'} - E_k^A - E_K) D_{kk}(t) \tag{3.91}$$

The total scattering effects in Eq. 3.39 are therefore given by the sum of Eqs. 3.80 and 3.91.

E. The Neutron Balance Equation in Continuous Momentum Space

The neutron balance equation, as given in Eq. 3.39 has been reduced by a systematic study of the various collision terms. Expressions for the associated transition probabilities, although still rather formal, have been derived using Heitler's damping theory. Having determined the explicit dependence upon neutron occupation number of each process, we thus obtain an equation describing the neutron density $F(\mathbf{X}, \mathbf{K}, t)$. This equation can be written as

$$\left[\frac{\partial}{\partial t} + \frac{\hbar}{m} \mathbf{K}_j \frac{\partial}{\partial X_j} + R_T(\mathbf{X}, \mathbf{K})\right] F(\mathbf{X}, \mathbf{K}, t)$$

$$= \sum F(\mathbf{X}, \mathbf{K}', t) [R_s(\mathbf{X}; \mathbf{K}' \to \mathbf{K}) + R_F(\mathbf{X}; \mathbf{K}' \to \mathbf{K})] \tag{3.92}$$

where we have introduced the following reaction rates,

$$R_T(x) = R_c(x) + R_s(x) + R_F(x) \tag{3.93}$$

$$R_c(\mathbf{X}, \mathbf{K}) = \sum_{lk'\varkappa'\lambda's} [1 + F_{\lambda'}(\mathbf{X}, \boldsymbol{\kappa}', t)] \tfrac{1}{2} r_{k'\mathbf{K}'\lambda', k\mathbf{K}s}^{cl} \times$$

$$\times D_{kk}(t) \delta(E_{k'}^{A+1} + E_{\varkappa'} - B^{A+1} - E_k^A - E_K) \tag{3.94}$$

$$R_s(\mathbf{X}, \mathbf{K}) = \sum_{lk'\mathbf{K}'s'ks} \left[1 - \frac{V}{2} F(\mathbf{X}, \mathbf{K}', t)\, D_{kk}(t)\right] \times$$
$$\times \tfrac{1}{2} \Big\{ r^{Es}_{k'\mathbf{K}'s', k\mathbf{K}s}\, \delta(E^A_k + E_K - E^{A'}_{k'} - E_{K'})$$
$$+ \sum_{\varkappa'\lambda'} [1 + F_{\lambda'}(\mathbf{X}, \boldsymbol{\kappa}', t)]\, r^{Isl}_{k'\mathbf{K}'\varkappa'\lambda', k\mathbf{K}s} \times$$
$$\times\ \delta(E^A_{k'} + E_{K'} + E_\varkappa - E^A_k - E_K) \Big\} \tag{3.95}$$

$$R_F(\mathbf{X}, \mathbf{K}) = \sum_{\substack{lJk_1\alpha_1\\ ks\{K\}'KJ}} \prod^*_{\{K'\}JK} \left[1 - \frac{V}{2} F(\mathbf{X}, \mathbf{K}_j, t)\, D_{kk}(t)\right] \times$$
$$\times \tfrac{1}{2}\, r^{Fl}_{k_1\alpha_1, k\mathbf{K}s}(\{K\}'_{JK})\, \delta(E^A_k + E_K - E_f) \tag{3.96}$$

$$R_s(\mathbf{X}; \mathbf{K}' \to \mathbf{K}) = \left[1 - \frac{V}{2} F(\mathbf{X}, \mathbf{K}, t)\right] \sum_{lk's'ks} D_{k'k'}(t) \times$$
$$\times \tfrac{1}{2} \{ r^{Es}_{k\mathbf{K}s, k'\mathbf{K}'s'}\, \delta(E^A_k + E_K - E^A_{k'} - E_{K'})$$
$$+ \sum_{\varkappa, \lambda} [1 + F_\lambda(\mathbf{X}, \varkappa, t)]\, r^{Isl}_{k\mathbf{K}s\varkappa\lambda, k'\mathbf{K}'s'} \times$$
$$\times\ \delta(E^A_{k'} + E_{K'} - E^A_k - E_K - E_\varkappa) \} \tag{3.97}$$

$$R_F(\mathbf{X}; \mathbf{K}' \to \mathbf{K}) = \sum_{\substack{lJk_1\alpha_1\\ k's'\{K\}JK}} \prod^*_{\{K\}JK} \left[1 - \frac{V}{2} F(\mathbf{X}, \mathbf{K}_j, t)\right] D_{k'k'}(t) \times$$
$$\times \tfrac{1}{2}\, r^{Fl}_{k_1\alpha_1, k'\mathbf{K}'s'}(\{K\}_{JK})\, \delta(E^A_{k'} + E_{K'} - E_f) \tag{3.98}$$

The expressions for these reaction rates have been discussed in the previous sections in this chapter. Eq. 3.92 is now seen to be identical in structure to the conventional neutron transport Eq. 1.1; however, the present equation has been derived on the basis of a discrete phase space.

It has been indicated earlier that the distribution of momentum points is so dense compared to resolutions in any practical measurement that no appreciable error can result by expressing Eq. 3.92 as an equation in continuous neutron momentum space. This is readily accomplished by the use of Eq. 2.30 whereupon we find

$$\left[\frac{\partial}{\partial t} + \mathbf{v} \cdot \nabla + v\Sigma_t(\mathbf{X}, \mathbf{v})\right] f(\mathbf{X}, \mathbf{v}, t)$$
$$= \int d^3v'\, v'\, f(\mathbf{X}, \mathbf{v}', t)\, [\Sigma_s(\mathbf{X}, \mathbf{v}')\, \mathscr{F}(\mathbf{v}' \to \mathbf{v}) + \Sigma_F(\mathbf{X}, \mathbf{v}')\, \mathscr{L}(\mathbf{v}' \to \mathbf{v})] \tag{3.99}$$

where we have introduced the neutron velocity as a variable, $\mathbf{v} = \hbar\mathbf{K}/m$, and have expressed the collision rates in terms of corresponding macroscopic cross sections, Σ_i. The two frequencies, $\mathscr{F}$ and $\mathscr{L}$ in Eq. 3.99, are defined as follows:

$$\begin{aligned}\sum_{\mathbf{K}\in d^3K} R_s(\mathbf{X};\mathbf{K}' \to \mathbf{K}) &= R_s(\mathbf{X},\mathbf{K}')\,\mathscr{F}(\mathbf{K}' \to \mathbf{K})\,d^3K \\ &= v'\,\Sigma_s(\mathbf{X},\mathbf{v}')\,\mathscr{F}(\mathbf{v}' \to \mathbf{v})\,d^3v \end{aligned} \tag{3.100}$$

$$\begin{aligned}\sum_{\mathbf{K}\in d^3K} R_F(\mathbf{X};\mathbf{K}' \to \mathbf{K}) &= R_F(\mathbf{X},\mathbf{K})\,\mathscr{L}(\mathbf{K}' \to \mathbf{K})\,d^3K \\ &= v'\Sigma_F(\mathbf{X},\mathbf{v}')\,\mathscr{L}(\mathbf{v}' \to \mathbf{v})\,d^3v \end{aligned} \tag{3.101}$$

As usual, $\mathscr{F}(\mathbf{v}' \to \mathbf{v})\,d^3v$ is to be interpreted as the probability that a given neutron scattered with velocity $\mathbf{v}'$ will have its final velocity in d^3v about $\mathbf{v}$, whereas $\mathscr{L}(\mathbf{v}' \to \mathbf{v})\,d^3v$ represents the expected number of neutrons emitted with velocity in d^3v about $\mathbf{v}$, providing that a fission event has been initiated by a neutron with velocity $\mathbf{v}'$.

For the reactions of interest the cross sections and $\mathscr{L}$ are independent of the direction of the incident neutron. Moreover, $\mathscr{F}$ often* depends only upon the initial and final speeds and the scattering angle, $\theta = \cos^{-1}(\mathbf{v}\cdot\mathbf{v}'/|\mathbf{v}|\,|\mathbf{v}'|)$. Inserting these simplifications into Eq. 3.99 and assuming that the discrete configuration space can be replaced by a continuum, we finally obtain the neutron transport equation in a form that is conventionally employed in all investigations of neutron slowing down, diffusion, and thermalization.

References

1. J. M. Blatt and V. F. Weisskopf, *Theoretical Nuclear Physics*, John Wiley Sons Inc., New York, 1952.
2. A. M. Lane and R. G. Thomas, *Rev. Mod. Phys.*, **30**: 257 (1958).
3. W. Heitler, *The Quantum Theory of Radiation*, Oxford University Press, New York, 1954, third edition.
4. E. Arnous and W. Heitler, *Proc. Royal Soc.*, **220A**: 290 (1953).
5. R. C. O'Rourke, "Damping Theory", Naval Research Laboratory Report 5315, 1959.
6. A. M. Weinberg and E. P. Wigner, *The Physical Theory of Neutron Chain Reactors*, University of Chicago Press, Chicago, Ill., 1958.

---

* An exception might be the scattering frequency for low-energy neutrons in crystals.

7. E.H.Klevans, Thesis, University of Michigan, Ann Arbor, Michigan, 1962.
8. A.Z.Akcasu, Thesis, University of Michigan, Ann Arbor, Michigan, 1963.
9. A.Z.Akcasu, University of Michigan Technical Report 04836-1-T, April 1963.
10. R.V.Churchill, *Operational Mathematics*, McGraw-Hill Book Company, Inc., New York, 1958.
11. L.I.Schiff, *Quantum Mechanics*, McGraw-Hill Book Company, Inc., New York, 1955, second edition.
12. L.L.Foldy, *Phys. Rev.*, **87**: 693 (1952); *Rev. Mod. Phys.*, **30**: 471 (1958).
13. L.Dresner, "Resonance Absorptions of Neutrons in Nuclear Reactors", ORNL-2659 (1959); *Resonance Absorption in Nuclear Reactors*, Pergamon Press, New York, 1960.
14. L.Nordheim, "Theory of Resonance Absorption", GA-638 (1959).
15. J.B.Sampson and J.Chernick, *Prog. in Nuclear Energy*, Series I, **223** (1958).
16. S.Oleksa, *Proceedings of Brookhaven Conference on Resonance Absorption in Nuclear Reactors*, BNL–433, 59 (1958).
17. R.E.Marshak, *Rev. Mod. Phys.*, **19**: 185 (1947).
18. J.H.Ferziger and P.F.Zweifel, *The Slowing Down of Neutrons in Nuclear Reactors,* Pergamon Press (to be published in 1966).
19. D.J.Hughes and R.B.Schwartz, *Neutron Cross Sections*, BNL-325 (1958), second edition.
20. E.Fermi, *Ricerca Scientifica*, **7**: 13 (1936); English translation available as USAEC Rept. NP-2385.
21. G.Breit, *Phys. Rev.*, **71**: 215 (1947).
22. J.P.Plummer and G.C.Summerfield, *Phys. Rev.* **131**, 1153 (1963).
23. G.C.Summerfield, *Ann. Phys.*, **26**, 72 (1964).
24. D.M.Chase, L.Wilets, and A.R.Edmonds, *Phys. Rev.*, **110**: 1080 (1958).

IV

Neutron-nuclear Interactions: Medium Effects

In detailed investigations of neutron transport in macroscopic systems, the use of adequate cross sections in the transport equation is essential. And adequacy here requires that the cross sections not only reflect the specifically nuclear processes under consideration, but also all relevant environmental effects. The environment can significantly influence the description of the cross section in at least two ways. The dynamics and symmetries of the system can either separately or simultaneously modify an observed reaction rate.

The ratio of nuclear force ranges to characteristic internuclear distances is of the order of 10^{-5} or less. Thus it is anticipated that a given neutron will interact with the nuclei in any medium one at a time. Nevertheless the probability of a collision between a neutron and a nucleus will be affected (because of the requirements of energy and momentum conservation) by the character of the states available to the target nucleus in the system. In turn, the nature of these states is determined by the dynamics of the macroscopic system. Furthermore, system dynamics modifies reaction rates in still another way, since they will depend upon the relative probabilities of finding a target nucleus in particular available states before a collision occurs. The effects on reaction rates depending upon the nature of the available states for the nuclei are often referred to as "binding effects", whereas those depending upon the probabilities of occupancy of these states are called "Doppler effects".

System symmetries, which for practical purposes may be regarded as distinct from system dynamics, can also play a role in determining reaction rates for neutrons at sufficiently low energies that their de

Broglie wavelengths approach or exceed internuclear spacings, i.e., energies of the order of tenths of an electron volt or less. The most striking example of symmetry effects on neutron cross sections is probably Bragg scattering in crystals.

For the very low-energy neutrons for which symmetry effects markedly influence reaction rates, dynamical effects of both kinds (binding and Doppler) are generally expected to be significant also. Since molecular dissociation and crystal dislocation potential energies are typically of the order of a few electron volts, it is anticipated that, at least in principle, there will be neutron reaction rates that are affected by both aspects of system dynamics, but not by symmetries. Finally, for still higher-energy neutrons, binding effects should decrease in importance and only the Doppler effect should remain as an influence on cross sections.

The expressions for neutron-nuclear reaction rates that have been derived in the previous chapter implicitly include all of these effects. In this chapter we shall explicitly investigate some aspects of them.

The following discussion is restricted to radiative capture and elastic scattering because, for simple systems, the calculation involved is straightforward and the results obtained are of considerable interest from the standpoint of reactor analysis.*

Because of the complexities of inelastic scattering and fission reactions and of our intention to describe them only qualitatively, a quantitative investigation of medium effects in these processes does not seem feasible at this point. Furthermore, it is unlikely that inelastic scattering reactions will be observably sensitive to medium properties due to the large neutron energy required. It is also unlikely that fission reactions will be influenced by binding, although Doppler effects may be important.

One neutron-nuclear reaction in which medium effects are prominent is elastic potential scattering at low energies. With the advent of high-flux reactors and the development of high-resolution neutron spectrometry, it has become feasible to measure in considerable detail the energy and angular distributions of the scattered neutrons. These investigations not only provide cross section data for reactor calculations,

---

* For example, the temperature dependence of reactivity[1] and the study of thermalization and diffusion of neutrons in the energy region[2] below 1 eV are of interest.

but also constitute a quantitative means of studying atomic motions in solids and liquids.*

In the latter cases the emphasis is on the proper interpretation of the measurements, and for this purpose a realistic description of the scattering system must exist. The theory of neutron scattering by crystals and low-density gases, on the basis of available models capable of representing quite accurately the motions in actual systems, has been developed to the extent that quantitative understanding of the various processes involved is possible. On the other hand, the corresponding theory for liquids (and, in fact, the theory of liquid state in general) suffers from the lack of a systematic and reliable description of molecular motions. The complexities of these motions make it necessary to introduce simplifying assumptions and specialized models to carry out an analysis. Since much of the theory of neutron scattering by liquids is still under development, this aspect of the investigation of medium effects will not be considered in the present work.

From the reaction-rate expressions already derived it can be seen that the effects of the external degrees of freedom of the nuclei are partially specified by the matrix elements of $\exp(i\mathbf{q}\cdot\mathbf{r})$, where $\mathbf{q}$ is either a neutron or photon wave vector. The direct calculation of these matrix elements necessarily involves specification of dynamical and symmetry properties of the medium. For the purpose of illustration we shall consider two simple systems, the ideal gas and the Einstein crystal. Although these are rather idealized descriptions of actual systems, they are capable of providing useful cross sections for realistic reactor calculations. The discussion presented here is primarily concerned with the translational motions of the atoms. Thus the results obtained, strictly speaking, are applicable only to monatomic systems. For polyatomic molecules the same procedure may be used to treat the center-of-mass degrees of freedom, but in addition molecular rotations and internuclear vibrations must also be taken into account.

---

* For an extensive list of references as well as a number of important papers see the proceedings of the "Symposium on Inelastic Scattering of Neutrons in Solids and Liquids" held in Vienna, 1960, and the proceedings of a similar symposium held in Chalk River, and in Bombay, 1964.[3]

A. Transport in an Ideal Gas

The simplest dynamical system appropriate for the discussion of medium effects is one in which the atoms do not interact appreciably with each other. The recoil of these atoms in a collision with neutrons will be like that of free particles, so that only two properties of the system can be expected to influence the cross sections: the particle mass (which influences the magnitude of the recoil) and the temperature (which characterizes the average energy of the atoms). The use of a free-particle description makes the cross-section calculations quite easy, and the results are often useful in transport problems, since systems other than dilute gases can also be treated in this manner whenever neutron energies are such that chemical binding can be ignored.

Radiative Capture

The reaction rate describing radiative capture in isolated levels has been derived in the preceding chapter. From Eqs. 3.94 and 3.59 we have

$$R_c = \sum_{\substack{lk'\varkappa\lambda' \\ ks\alpha}} [1 + F_{\lambda'}(\mathbf{X}, \boldsymbol{\kappa}', t)]\, \delta(E_{k'}^{A+1} - B^{A+1} + E_{\varkappa'} - E_k^A - E_K) \times$$
$$\times\, |U_{0\alpha}^{Rl}(\boldsymbol{\kappa}'\lambda')|^2\, |U_{\alpha 0}^{Nl}(Ks)|^2 \frac{\pi}{\hbar} D_{kk}(t) \times$$
$$\times \left| \sum_{k''} \frac{\langle k'|\, e^{-i\varkappa'\cdot\mathbf{R}_l}\, |k''\rangle \langle k''|\, e^{i\mathbf{K}\cdot\mathbf{R}_l}\, |k\rangle}{\mathscr{E}_\alpha + E_{k''}^{A+1} - E_k^A - E_K - \dfrac{i}{2}\Gamma_\alpha} \right|^2 \tag{4.1}$$

For a system of noninteracting particles with no internal degrees of freedom the Hamiltonian is simply the sum of individual particle kinetic energies. The kth eigenstate of the system is a product of individual-particle cell eigenstates, each characterized by a wave vector which is discrete in exactly the same sense as for the neutrons. The two matrix elements in the k'' sum can now be written as

$$\langle k'|\, e^{-i\varkappa'\cdot\mathbf{R}_l}\, |k''\rangle \langle k''|\, e^{i\mathbf{K}\cdot\mathbf{R}_l}\, |k\rangle = \delta_K(\mathbf{k}_l' + \boldsymbol{\kappa}' - \mathbf{k}_l'')\, \delta_K(\mathbf{k}_l'' - \mathbf{K} - \mathbf{k}_l) \tag{4.2}$$

where $\mathbf{k}_l$, $\mathbf{k}_l''$, and $\mathbf{k}_l'$ are wave vectors characterizing the lth nucleus in initial (mass A), intermediate, and final (mass $(A + 1)$) states respectively. The symbol $\delta_K(\mathbf{x} - \mathbf{x}')$ denotes a Kronecker delta. Because of

Eq. 4.2 the absolute square of the k'' sum becomes

$$\frac{\delta_K(\mathbf{k}' + \boldsymbol{\kappa}' - \mathbf{K} - \mathbf{k})}{(\mathscr{E}_\alpha - E_{Kk})^2 + (\Gamma_\alpha/2)^2} \tag{4.3}$$

where

$$E_{Kk} = \frac{\mu}{2}\left(\frac{\hbar\mathbf{K}}{m} - \frac{\hbar\mathbf{k}}{M}\right)^2 \tag{4.4}$$

$\mu = mM/(m + M)$ and we have suppressed the subscript l. As one may expect, in a collision in which the nucleus is in motion the effective neutron energy is the relative energy E_{Kk}.

In the energy-conserving delta function in R_c the molecular energies ($E_k = \hbar^2k^2/2M$) are exceedingly small compared to the binding energy B^{A+1} or the photon energy $E_{\varkappa'}$. If we assume that the difference $E_{k'}^{A+1} - E_k^A$ can be effectively ignored,* the summation over k' states can be performed immediately since then only the Kronecker delta depends upon $\mathbf{k}'$. The energy of the photon emitted during the capture process is usually many orders of magnitude greater than the characteristic thermal energy k_BT, where k_B is the Boltzmann constant and T the equilibrium temperature. Therefore it is usually justified to neglect in Eq. 4.1 the distribution $F_{\lambda'}(\mathbf{X}, \boldsymbol{\kappa}', t)$ compared to unity. We now have

$$R_c = \frac{\pi}{\hbar} \sum_{l\varkappa'\alpha} \delta(E_{\varkappa'} - B^{A+1} - E_K) \sum_{\lambda'} |U_{0\alpha}^{Rl}|^2 \sum_{s} |U_{\alpha 0}^{Nl}|^2 \times$$
$$\times \sum_{k} \frac{D_{kk}(t)}{(\mathscr{E}_\alpha - E_{Kk})^2 + (\Gamma_\alpha/2)^2} \tag{4.5}$$

It is convenient to replace the k sum by an appropriate integral. This is accomplished by letting the system volume become arbitrarily large and observing that

$$\sum_{\mathbf{k}\in \mathrm{d}^3k} D_{kk}(t) = P(\mathbf{k})\, \mathrm{d}^3k \tag{4.6}$$

In Eq. 4.6 it is often assumed that the system is in a thermodynamic state so that P is time-independent. The sum over photon momentum can also be replaced by an integral and in so doing we may introduce the radiation and neutron partial widths as given in Eqs. 3.61 and 3.63. The ratio

$$(E_\alpha^{A+1})^{-2} \int_0^\infty \mathrm{d}E_\varkappa\, E_\varkappa^2\, \delta(E_\varkappa + B^{A+1} - E_K) \tag{4.7}$$

* Keeping this difference entails no difficulty in principle. The approximation is made here for convenience in calculation.

is seen to be essentially unity in view of the neglect of molecular energies. After some simplification the absorption rate becomes

$$R_c = \frac{\pi\hbar}{2mKL^3} \sum_{l\alpha} \Gamma_\alpha^{(N)} \Gamma_\alpha^{(R)} \int \mathrm{d}^3k \frac{P(\mathbf{k})}{(\mathscr{E}_\alpha - E_{Kk})^2 + (\Gamma_\alpha/2)^2} \tag{4.8}$$

The microscopic cross section σ is related to the reaction rate R by dividing the latter by the incident neutron speed and the nuclear density. Since R_c represents the neutron absorption rate by mass A nuclei located in the configuration volume specified by $\mathbf{X}$, the l sum in Eq. 4.8 merely gives a factor of $N_A(\mathbf{X})$, where $N_A(\mathbf{X})$ is the total number of mass A nuclei in the cell. The nuclear density in this case is $N_A(\mathbf{X})\, L^{-3}$ so that

$$\begin{aligned} \sigma_c(K) &= [L^3/N_A(\mathbf{X})]\, \Sigma_c(\mathbf{X}, K) \\ &= \frac{\pi\lambda^2}{2} \sum_\alpha \Gamma_\alpha^{(N)} \Gamma_\alpha^{(R)} \int \mathrm{d}^3k \frac{P(\mathbf{k})}{(\mathscr{E}_\alpha - E_{Kk})^2 + (\Gamma_\alpha/2)^2} \end{aligned} \tag{4.9}$$

where Σ_c is the macroscopic capture cross section and $\lambda = 1/K$. For systems in a thermodynamic state we may use for $P(\mathbf{k})$ the Maxwell-Boltzmann distribution, and we then find that σ_c depends parametrically upon the medium temperature. Eq. 4.9 therefore gives the familiar single-level resonance capture cross section.*

The energy dependence of each term in Eq. 4.9 gives the so-called resonance line shape. In the limit of zero temperature $P(\mathbf{k})$ becomes $\delta(\mathbf{k})$, and

$$\sigma_c(K) = \frac{\pi\lambda^2}{2} \sum_\alpha \frac{\Gamma_\alpha^{(N)} \Gamma_\alpha^{(R)}}{(\mathscr{E}_\alpha - E_{K0})^2 + (\Gamma_\alpha/2)^2} \tag{4.10}$$

describes radiative capture by a stationary absorber. Note however, Eq. 4.10 still contains the effect of recoil of the compound nucleus. Each line shape in this case is called "natural", the Lorentzian being characterized by a width $\Gamma_\alpha/2$. At finite temperatures, the integral (4.9) gives a weighted superposition of many Lorentzians so the resulting line shape can be significantly broadened, but with an accompanying depression of the peak value. This effect is known as "Doppler broadening" and is of considerable importance in studies of reactor safety and control,†

* The effect of thermal motion upon radiative capture of neutrons by gas-phase nuclei was first considered by Bethe and Placzek.[4]

† For a review of Doppler effect in thermal reactors see Pearce.[1] The effect in fast reactors has been discussed by Feshbach *et al.*[5] and by Nicholson.[1] Recently the problem of nonuniform temperature distribution has been investigated by Olhoeft.[6]

for it is well known that the broadening of a resonance line can cause a significant increase in the effective absorption in a system.

The integral in Eq. 4.9 can be reduced to a form that is conventional in the investigation of Doppler effect in reactors.* In terms of velocity variables,

$$\sigma_c(v) = \frac{\pi \lambda^2}{2} \sum_\alpha \Gamma_\alpha^{(N)} \Gamma_\alpha^{(R)} \int d^3V \frac{P(V)}{(\mathscr{E}_\alpha - E_r)^2 + (\Gamma_\alpha/2)^2} \tag{4.11}$$

where

$$P(V) = \left(\frac{M}{2\pi k_B T}\right)^{3/2} e^{-MV^2/2k_BT}$$

$$E_r = \mu v_r^2/2$$

$$\mathbf{v}_r = \mathbf{v} - \mathbf{V},$$

$$\mathbf{v} = \hbar \mathbf{K}/m$$

$$\mathbf{V} = \hbar \mathbf{k}/M \tag{4.12}$$

Since $\mathbf{v}$ is a fixed vector in the integration, the integral becomes

$$2\pi \int_0^\infty V \, dV \, P(V) \int_{|v-V|}^{v+V} dv_r \frac{v_r}{(\mathscr{E}_\alpha - E_r)^2 + (\Gamma_\alpha/2)^2}$$

By changing the order of integration and performing the V integral, we obtain

$$\sigma_c(v) = 4\pi\lambda^2 \sum_\alpha \Gamma_\alpha^{(N)} \Gamma_\alpha^{(R)} \Gamma_\alpha^{-2} \psi(\xi_\alpha, x_\alpha) \tag{4.13}$$

$$\psi(\xi, x) = \frac{\xi}{2\sqrt{\pi}} \int_{-\infty}^{\infty} dy \frac{e^{-\xi^2(x-y)^2/4}}{1 + y^2} \tag{4.14}$$

where we have introduced the variables

$$x_\alpha = 2(E - \mathscr{E}_\alpha)/\Gamma_\alpha$$

$$y = 2(\mathscr{E}_\alpha - E_r)/\Gamma_\alpha \tag{4.15}$$

$$\Delta^2 = 4m E k_B T/M$$

$$\xi_\alpha = \Gamma_\alpha/\Delta$$

* See, for example, L. W. Nordheim, "Resonance Absorption of Neutrons". Lectures at the Mackinac Island Conference on Neutron Physics, June, 1961, available as a report of the Michigan Memorial Phoenix Project, University of Michigan.

In arriving at this result it has been assumed that $\mu \approx m$ and that in the exponential

$$\sqrt{E_r} \approx \sqrt{E}\left(1 + \frac{E_r - E}{2E}\right)$$

The integral ψ has been studied extensively[7] and its values as a function of ξ and x are tabulated.[8] It is somewhat interesting to note that at very high temperatures (ξ small) the contribution to the integral comes mainly from $y = 0$. The resonance line shape is then essentially governed by the Gaussian $\exp(-x^2\xi^2/4)$, the width of which, $2(4Ek_BT/A)^{1/2}$, is known as the Doppler width. The parameter $\xi/2$ therefore is the ratio of natural width to Doppler width.

Elastic Scattering

From the preceding section it is observed that the external degrees of freedom of nuclei in gases influence a given collision only kinematically. Because not all atoms move with the same velocity, the cross section appears as an average over a distribution (usually thermodynamic) of target velocities. The same remarks are also applicable to elastic scattering, and in the case of potential scattering the average is rather easily performed. The reaction rate describing an elastic process in which the neutrons suffer an energy change of $E_{K'} - E_K$ and direction change of $\theta = \cos^{-1}(\hat{\mathbf{K}} \cdot \hat{\mathbf{K}}')$ is given by

$$R_s = \left(\frac{L}{2\pi}\right)^3 \frac{1}{2} \sum_{kk'} \delta(E_k^A - E_{k'}^A + E_K - E_{K'})\, P_{k'}(t) \sum_{ss'} \Theta \qquad (4.16)$$

where

$$\Theta = \frac{2\pi}{\hbar}\left|\left[\sum_l \langle k|\, e^{-i\mathbf{Q}\cdot\mathbf{R}_l}\, U_l |\, k'\rangle - \sum_{\alpha k''} \frac{U_{0\alpha}^{Nl}(Ks)\, U_{\alpha 0}^{Nl}(K's')\, \langle k|\, e^{-i\mathbf{K}\cdot\mathbf{R}_l}\, |k''\rangle\, \langle k''|\, e^{i\mathbf{K}'\cdot\mathbf{R}_l}\, |k'\rangle}{\mathscr{E}_\alpha + E_{k''}^{A+1} - E_{k'}^A - E_{K'} - \frac{i}{2}\Gamma_\alpha}\right]\right|^2 \qquad (4.17)$$

in which we have replaced $D_{kk}(t)$ by $P_k(t)$ as in Eq. 4.6.

In calculating the various matrix elements, we note from Eq. 3.75 in the expression for U_l that the integrand contains the step function

$E(\mathbf{X}, \mathbf{R} + \mathbf{R}_l)$ as well as v_i^l. Because of the short range of nuclear forces $L \gg \lambda_i$, and we may effectively write the step function as $E(\mathbf{X}, \mathbf{R})$ and obtain

$$\langle k| \, U_l \, \mathrm{e}^{-i\mathbf{Q}\cdot\mathbf{R}_l} \, |k'\rangle = \overline{U}_l \, \delta_K(\mathbf{k} - \mathbf{k}' + \mathbf{Q}) \tag{4.18}$$

where

$$\overline{U}_l \approx L^{-3} \int \mathrm{d}^3R \, \mathrm{e}^{-i\mathbf{Q}\cdot\mathbf{R}} \, [\delta_{ss'} v_0^l(R) + \mathbf{I}_l \cdot u_j^+(s) \, \sigma_{jk} u_k(s') \, v_1^l(R)] \tag{4.19}$$

Again, the subscript l appropriate to the nuclear momenta in the Kronecker delta is understood. For potentials which depend only upon the magnitude of $\mathbf{R}$ (as assumed here) $\overline{U}_l$ is real. The matrix elements in Θ which describe resonant scattering are given by Eq. 4.2 with $\mathbf{K}'$ replacing $\varkappa'$, so the k'' sum can be treated as before. The momentum-conserving Kronecker deltas appearing in both terms of Θ involve only the neutron and the l nucleus,* the sum is therefore incoherent and may be removed outside the square of the absolute value. This sum again gives a factor of N_A. If we further assume that the resonances do not overlap,

$$\Theta = \frac{2\pi N_A}{\hbar} \, \delta_K(\mathbf{k} - \mathbf{k}' + \mathbf{Q}) \times$$

$$\times \left[\overline{U}^2 + \sum_\alpha \frac{|U_{0\alpha}^N|^2 \, |U_{\alpha 0}^N|^2 - 2\overline{U}(\mathscr{E}_\alpha - E_{K'k'}) \, |U_{0\alpha}^N|^2}{(\mathscr{E}_\alpha - R_{K'k'})^2 + (\Gamma_\alpha/2)^2} \right] \tag{4.20}$$

In writing the cross terms in Eq. 4.20 which represent the interference between potential and resonant scatterings, it has been assumed that the neutron emission and absorption matrix elements are at most only weakly dependent upon momentum and spin so that

$$U_{0\alpha}^N(Ks) \, U_{\alpha 0}^N(K's') \approx |U_{0\alpha}^N|^2$$

This approximation eliminates the explicit occurrence of real terms proportional to i. The particular model describing potential scattering used here has been introduced with a spin-dependent term. Spin effects can also be taken into account in the analysis of resonant scattering, although this particular aspect has not been emphasized. In the interest of illustrating the dynamical consequences of macroscopic medium effects we shall ignore the effects arising from neutron-nuclear spin

---

* This is only true for ideal gases in which there is no interparticle interaction.

coupling in our discussion. This neglect implies the following:

$$\sum_{ss'} \bar{U}^2 = 2\bar{U}^2,$$

$$\sum_{ss'} |U^N_{0\alpha}(Ks)|^2 \, |U^N_{\alpha 0}(K's')|^2 \approx \left(\frac{\pi\hbar^2}{mL^3} \, \lambda\Gamma^{(N)}_\alpha\right)^2 \tag{4.21}$$

$$\sum_{ss'} \bar{U}|U^N_{0\alpha}|^2 \approx \bar{U}\left(\frac{\pi\hbar^2}{mL^3} \, \lambda\Gamma^{(N)}_\alpha\right)$$

Making use of these results and inserting Eq. 4.20 into R_s, we obtain

$$R_s = \left(\frac{L}{2\pi}\right)^3 \frac{2\pi N_A}{\hbar} \sum_{kk'} \delta(E_k - E_{k'} + \Delta E) \, P_{k'}(t) \, \delta_K(\mathbf{k} - \mathbf{k}' + \mathbf{Q}) \times$$

$$\times \left\{\bar{U}^2 + \sum_\alpha \frac{\frac{1}{2}\left[\frac{\pi\hbar^2}{mL^3} \, \lambda\Gamma^{(N)}_\alpha\right]^2 - \bar{U}(\mathscr{E}_\alpha - E_{K'k'})\left[\frac{\pi\hbar^2}{mL^3} \, \lambda\Gamma^{(N)}_\alpha\right]}{(\mathscr{E}_\alpha - E_{K'k'})^2 + (\Gamma_\alpha/2)^2}\right\} \tag{4.22}$$

where $\Delta E = E_K - E_{K'}$, and we further suppress the superscript A in the energy symbols. At this point it becomes convenient to treat the neutron momentum as a continuous variable, then δ_K becomes a Dirac delta,

$$\delta_K(\mathbf{k} - \mathbf{k}' + \mathbf{Q}) = \left(\frac{2\pi}{L}\right)^3 \delta(\mathbf{k} - \mathbf{k}' + \mathbf{Q}) \tag{4.23}$$

Moreover, it is also appropriate to treat the k and k' sums as integrals.

In the case of spinless nuclei the potential $\bar{U}$ characterizing the direct process may be written as

$$\bar{U} = L^{-3} g \int d^3R \, \frac{e^{-i\mathbf{Q}\cdot\mathbf{R} - R/\lambda}}{R} \approx \frac{2\pi\hbar^2}{mL^3} \, \chi(Q) \tag{4.24}$$

with

$$\chi(Q) = 2gm \, \frac{(\lambda/\hbar)^2}{1 + (\lambda Q)^2} \tag{4.25}$$

Comparing with Eq. 3.75 we see that this expression is valid so long as $\lambda \ll L$, i.e., range of interaction small compared to the linear dimension of the spatial cell. Since λ is of order 10^{-13} cm, this condition is always fulfilled for any reasonable choise of L. If, on the other hand, we had used the Fermi pseudopotential instead of the Yukawa potential we would have obtained

$$\chi = \left(\frac{m + M}{M}\right) a \tag{4.26}$$

where χ is the bound-atom scattering length and a is the conventional "free-atom" scattering length.

The final form of the elastic-scattering reaction rate may now be expressed as

$$R_s = \frac{\hbar^3}{m^2} \int d^3k\, d^3k'\, f_A(\mathbf{X}, \mathbf{k}', t)\, \delta(E_k - E_{k'} + \Delta E)\, \delta(\mathbf{k} - \mathbf{k}' + \mathbf{Q}) \times$$

$$\times \left[\chi^2 + \sum_\alpha \frac{\frac{1}{2}(\lambda \Gamma_\alpha^{(N)}/2)^2 - \chi(\mathscr{E}_\alpha - E_{K'k'})\,(\lambda \Gamma_\alpha^{(N)}/2)}{(\mathscr{E}_\alpha - E_{K'k'})^2 + (\Gamma_\alpha/2)^2}\right] \tag{4.27}$$

Here we have denoted $[N_A(\mathbf{X})/L^3]\, P(\mathbf{k}', t)$ by $f_A(\mathbf{X}, \mathbf{k}', t)$ which, as an analogue of the neutron density, represents the average number of nuclei of mass A having momentum $\hbar\mathbf{k}'$ to be found at $\mathbf{X}$ at time t. For $\chi > 0$, the last term in Eq. 4.27 shows that the interference between potential and resonant scatterings is destructive at the low-energy side and constructive at the high-energy side of the resonance line.

The results in the present section can be summarized in terms of an equation describing the transport of neutrons in monatomic gases in which the dominant neutron-nuclear interactions are radiative capture and elastic scattering,

$$\left[\frac{\partial}{\partial t} + \frac{\hbar \mathbf{K}}{m} \cdot \nabla + \frac{\hbar K}{m} \Sigma_t(K)\right] f(\mathbf{X}, \mathbf{K}, t)$$

$$= \int d^3K'\, f(\mathbf{X}, \mathbf{K}', t) \left(\frac{\hbar K'}{m}\right) \Sigma_s(K')\, \mathscr{F}(K' \to \mathbf{K}) \tag{4.28}$$

where the macroscopic cross sections are

$$\Sigma_t(K) = \Sigma_c(K) + \Sigma_s(K), \tag{4.29}$$

$$\Sigma_c(K) = \frac{\pi}{4K^2} \sum_\alpha \Gamma_\alpha^{(N)} \Gamma_\alpha^{(R)} \int d^3k \frac{f_A(\mathbf{X}, \mathbf{k}, t)}{(\mathscr{E}_\alpha - E_{Kk})^2 + (\Gamma_\alpha/2)^2} \tag{4.30}$$

$$\Sigma_s(K) = \frac{\hbar^2}{mK} \int d^3k\, d^3k'\, d^3K'\, f_A(\mathbf{X}, \mathbf{k}, t)\, \delta(E_k - E_{k'} + \Delta E) \times$$

$$\times\, \delta(\mathbf{k} - \mathbf{k}' + \mathbf{Q}) \times$$

$$\times \left[\chi^2 + \sum_\alpha \frac{\frac{1}{2}(\lambda \Gamma_\alpha^{(N)}/2)^2 - \chi(\mathscr{E}_\alpha - E_{Kk})\,(\lambda \Gamma_\alpha^{(N)}/2)}{(\mathscr{E}_\alpha - E_{Kk})^2 + (\Gamma_\alpha/2)^2}\right] \tag{4.31}$$

and the scattering frequency $\mathscr{F}(\mathbf{K}' \to \mathbf{K})$ is given by

$$\Sigma_s(K')\,\mathscr{F}(\mathbf{K}' \to \mathbf{K}) = \frac{\hbar^2}{mK'}\int \mathrm{d}^3k\,\mathrm{d}^3k'\,f_A(\mathbf{X}, \mathbf{k}', t)\,\delta(E_k - E_{k'} + \Delta E) \times$$
$$\times\,\delta(\mathbf{k} - \mathbf{k}' + \mathbf{Q}) \times$$
$$\times\left[\chi^2 + \sum_\alpha \frac{\frac{1}{2}(\lambda\Gamma_\alpha^{(N)}2)^2 - \chi(\mathscr{E}_\alpha - E_{K'k'})\,(\lambda\Gamma_\alpha^{(N)}/2)}{(\mathscr{E}_\alpha - E_{K'k'})^2 + (\Gamma_\alpha/2)^2}\right] \qquad (4.32)$$

The cross sections, through their dependence upon the nuclear density, are of course also functions of position. The transport Eq. 4.28 has been extensively employed in reactor analysis. Usually, additional simplifications are introduced in attempts to treat a problem analytically. For example, in neutron thermalization and diffusion investigations it is conventional to assume "$1/v$" absorption* and neglect the effects of resonant scattering.

The potential part of the scattering cross section can be further reduced. Ignoring the contributions from resonant and interference effects we have from Eq. 4.31

$$\Sigma_s(K) = \frac{(\hbar\chi)^2}{mK}\int \mathrm{d}^3K'\,\mathrm{d}^3k\,f(\mathbf{X}, \mathbf{k}, t)\,\delta\left(E_R + \frac{\hbar}{M}\mathbf{Q}\cdot\mathbf{k} - \Delta E\right) \qquad (4.33)$$

where $E_R = (\hbar Q)^2/2M$ is the recoil energy and we have suppressed the nuclear mass designation in $f(\mathbf{X}, \mathbf{k}, t)$. Upon the use of the integral representation of the delta function this expression becomes

$$\Sigma_s(E) = \frac{\chi^2}{2\pi\hbar}\int_0^\infty \mathrm{d}E'\left(\frac{E'}{E}\right)^{1/2}\int \mathrm{d}\Omega'\int_{-\infty}^{\infty}\mathrm{d}t'\langle\ \rangle_T\,\mathrm{e}^{-\mathrm{i}t'(E_R - \Delta E)/\hbar} \qquad (4.34)$$

where

$$\langle\ \rangle_T = \int \mathrm{d}^3k\,f(\mathbf{X}, \mathbf{k}, t)\,\mathrm{e}^{-\mathrm{i}t'\hbar\mathbf{Q}\cdot\mathbf{k}/M} = n\,\mathrm{e}^{-k_BT(Qt')^2/2M} \qquad (4.35)$$

In Eq. 4.35 a Maxwellian distribution is again assumed in evaluating the integral. The nuclear density $[N(\mathbf{X})/L^3]$ is denoted here simply as n. The scattering cross section is now expressed in terms of the more conventional energy variable. From Eq. 4.34 we can identify a microscopic cross section $\sigma(E \to E', \theta)$ which describes the scattering of a neutron from energy E to E' with a specified change of direction of motion,

$$\Sigma_s(E) = n\int \mathrm{d}E'\,\mathrm{d}\Omega'\,\sigma(E \to E', \theta) \qquad (4.36)$$

---

* The inverse speed dependence of the capture cross section follows from Eq. 4.30 and 3.63 for neutron energies far below resonances.

where

$$\sigma(E \to E', \theta) = \chi^2 \left(\frac{E'}{E}\right)^{1/2} \frac{e^{-(E_R - \Delta E)^2/4E_R k_B T}}{(4\pi E_R k_B T)^{1/2}} \tag{4.37}$$

The variable θ denotes the angle between **K** and **K**$'$. The integration of σ over energy[9] or angles[10] can be carried out. These calculations, being quite complicated and not particularly illuminating, will not be discussed here. On the other hand, the high-energy limit is interesting and readily available. For $E \gtrsim 1$ eV the nuclei can be taken to be initially at rest,*

$$f(\mathbf{X}, \mathbf{k}, t) \approx n\delta(\mathbf{k})$$

and we obtain instead of Eq. 4.37

$$\sigma(E \to E', \theta) = \chi^2 \left(\frac{E'}{E}\right)^{1/2} \delta(E_R - \Delta E) \tag{4.38}$$

The angular integral of this result is readily performed,

$$\sigma(E \to E') = \frac{4\pi c^2}{E(1 - \alpha)} \frac{1}{1 + 2M\lambda^2(E - E')/\hbar^2}$$

when

$$\alpha E \leq E' \leq E; \tag{4.39}$$

otherwise,

$$\sigma(E \to E') = 0,$$

with $c = 2gm(\lambda/\hbar)^2$ and $\alpha = ((M - m)/(M + m))^2$. If the second factor is replaced by unity then $\sigma(E \to E')$ gives the scattering frequency familiar in reactor theory and leads to a constant total cross section $\sigma(E) = 4\pi c^2$.† This is equivalent to assuming that χ is a constant, which is actually valid for energies up to about 10^4 eV. Beyond this region experimental results show a gradual decrease in $\sigma(E)$ that can be fitted qualitatively by an expression of the form $(1 + \beta E)^{-1}$, β being an adjustable parameter. From Eq. 4.39 we find that $\sigma(E)$ is in fact given by this form,

$$\sigma(E) = \frac{4\pi c^2}{1 + 2M\lambda^2(1 - \alpha) E/\hbar^2}$$

* This is equivalent to the limit of zero temperature.

† The same result is obtained by using the Fermi pseudopotential instead of the Yukawa functions to describe neutron-nuclear interaction.

If resonant scattering is ignored, the transport Eq. 4.28, upon making use of the above results, becomes

$$\left[\frac{\partial}{\partial t} + v\mathbf{\Omega}\cdot\nabla + v\Sigma_t(v)\right]f(\mathbf{X}, \mathbf{v}, t)$$

$$= \int d^3v'\, v'\, f(\mathbf{X}, \mathbf{v}', t)\,\Sigma(v' \rightarrow v, \theta) \qquad (4.40)$$

with $\Sigma = n\sigma$. This integro-differential equation has been the fundamental equation in many investigations of neutron transport.[11] The energy-independent form of Eq. 4.40, the "one-speed" transport equation, provides a problem which can be treated with mathematical rigor.[12,13] The first two angular moments of Eq. 4.40 give the diffusion equation which constitutes the analytical basis for many of the present studies of nuclear reactors.

B. Transport in Crystals

It has been shown in the preceding section that, in the absence of chemical binding, nuclear recoil effects on the cross sections can sometimes be analyzed by a straightforward calculation. For systems in which interatomic forces cannot be ignored this effect is in general considerably more complicated. However, in the case of strong chemical binding, it is again possible to discuss medium effects in analytical terms, for then one can reasonably represent the atomic motions as oscillations and make use of well-developed dynamical models in solid-state theory.[14] After eliminating the dependence upon electronic coordinates[15] one obtains in the harmonic approximation, a description of nuclear motions identical to that for a set of coupled oscillators, which can then be decoupled by a transformation to normal coordinates. However it is our purpose to illustrate the general features of medium effects on neutron interactions. Thus to avoid an involved discussion of crystal dynamics we shall restrict our considerations to a system describable by a set of uncoupled oscillators. It will be assumed that each nucleus experiences identical interactions with its surroundings so that it executes isotropic oscillations about an equilibrium position (a lattice site in the crystal) independent of all other nuclei. From this it follows that the fundamental vibrational frequencies can all be taken to be the same.

The present model, the Einstein crystal, is admittedly a severe idealization of actual atomic motions in bound systems.* Nevertheless, the results derived on the basis of such a description are meaningful and, like the ideal-gas cross sections, often useful for practical calculations. In general it can be expected that the model will provide an adequate description of integral properties of the cross sections, but is not suitable for quantitative analysis of differential cross-section measurements. One can, however, extend the following results to more elaborate crystal descriptions; the required modifications being mainly refinements of the model and not changes in the method of calculation.

Radiative Capture

A natural extension of the investigation of neutron capture by free atoms[4] is the corresponding treatment for atoms bound in a crystal. This problem was first considered by Lamb[17] whose work is of more general significance because the process of radiative capture is closely related to other nuclear processes characterized by a point interaction, a fact that has attracted attention only recently.[18] For example, one encounters the same matrix elements of the form $\langle k| \exp (i\mathbf{q} \cdot \mathbf{R}) |k'\rangle$ in the problem of emission and absorption of nuclear gamma rays.† With respect to neutron absorption, Lamb showed that, if the lattice binding is sufficiently weak, the resonance line shape is the same as that of a free nucleus but at an effective temperature corresponding to the average energy per vibrational degree of freedom of the nucleus including zero-point vibration. We shall first obtain the general cross section and then show how this limit emerges.

The reaction rate to be studied is that given in Eq. 4.1. Since the energy eigenstates $|k\rangle$ are no longer momentum eigenstates, the matrix elements must be obtained by a different approach. We shall again ignore the external energies E_k^A and $E_{k'}^{A+1}$ in the delta function‡ and also

---

* In some crystals it is possible that the vibrational motions can be adequately described by an Einstein model. Such an example could be the hydrogen atom in zirconium hydride.[16]

† In fact, Lamb's theory provided the initial explanation of recoilless gamma-ray transitions, a process now known as the Mössbauer effect.[19]

‡ Although this approximation is conventional, it is here (as in Section A) not necessary. For example, the method of analysis used in the following to obtain elastic-scattering cross section can be equally well applied here.

neglect the photon distribution function compared to unity. The expression for R_c becomes

$$R_c = \frac{\pi}{\hbar} \sum_{l\varkappa'\lambda'\alpha s} \delta(E_{\varkappa'} - B^{A+1} - E_K)\, |U^{Rl}_{0\alpha}|^2\, |U^{Nl}_{\alpha 0}|^2 \langle\ \rangle \tag{4.41}$$

where

$$\langle\ \rangle = \sum_k P_k(t) \sum_{k''} \left| \frac{\langle k''|\, e^{i\mathbf{K}\cdot\mathbf{R}_l}\, |k\rangle}{\mathscr{E}_\alpha - E_K - E^A_k + E^{A+1}_{k''} - \frac{i}{2}\Gamma_\alpha} \right|^2 \tag{4.42}$$

where as a result of performing the k' summation the k'' sum appears outside the square of the modulus. The quantity $\langle\ \rangle$, except for a constant multiplicative factor, is identical to the expression considered by Lamb.[16] To carry out the k'' summation, we rearrange the resonance denominator by writing

$$(E - i\Gamma)^{-1} = i \int_0^\infty \mathrm{d}s\, e^{-(\Gamma - iE)s}$$

so that

$$\langle\ \rangle = \int \mathrm{d}s\, \mathrm{d}s'\, e^{-(s+s')\Gamma/2}\, e^{i(s-s')(E_K - \mathscr{E}_\alpha)} \langle\ \rangle_T \tag{4.43}$$

$$\langle\ \rangle_T = \sum_k P_k(t) \langle k|\, e^{i(s-s')H}\, e^{-i\mathbf{K}\cdot\mathbf{R}_l}\, e^{-i(s-s')H}\, e^{i\mathbf{K}\cdot\mathbf{R}_l}\, |k\rangle \tag{4.44}$$

For an Einstein crystal in thermal equilibrium it is a straightforward matter to evaluate $\langle\ \rangle_T$. The calculation is discussed in detail at the end of this section and we quote here only the result,

$$\langle\ \rangle_T = e^{-DK^2} \sum_{n=-\infty}^{\infty} I_n(PK^2)\, e^{-n[Z - i\hbar\omega(s-s')]} \tag{4.45}$$

where $D = \eta \coth Z$, $P = \eta \operatorname{csch} Z$, $\eta = \hbar/2M\omega$, $Z = \hbar\omega/2k_B T$, ω is the oscillator frequency and I_n the modified Bessel function. Inserting this result into Eq. 4.43 and performing the indicated integrals we obtain an expression for R_c. Upon introduction of the partial widths as before, the capture cross section is given by

$$\sigma_c(K) = \frac{\pi}{2K^2}\, e^{-DK^2} \sum_\alpha \Gamma^{(N)}_\alpha \Gamma^{(N)}_\alpha \sum_{n=-\infty}^{\infty} \frac{I_n(PK^2)\, e^{-nZ}}{(\mathscr{E}_\alpha - E_K - n\hbar\omega)^2 + (\Gamma_\alpha/2)^2} \tag{4.46}$$

The integer n denotes the number of phonons that are created or destroyed depending upon the sign of n. The case of $n = 0$ gives a resonance line centered at $\mathscr{E}_\alpha$. This absence of recoil corresponds to the fact that the neutron momentum is absorbed by the crystal as a whole. As in most crystal models, the mass of our system is assumed to be infinite. In practice, however, there will always be a finite, though vanishing small, amount of recoil. In general the line shape of each resonance can be quite complicated and may even show fine structure indicative of phonon transitions.[17]

It is of some interest to investigate the implications of Eq. 4.46 in the limits of strong and weak binding. The condition of tightly bound nuclei is simply expressed by taking the vibrational frequency to be arbitrarily large. In this limit the cross section becomes

$$\sigma_c(K) \underset{\omega\to\infty}{\approx} \frac{\pi}{2K^2} \sum \frac{\Gamma_\alpha^{(N)}\,\Gamma_\alpha^{(R)}}{(\mathscr{E}_\alpha - E_K)^2 + (\Gamma_\alpha/2)^2} \tag{4.47}$$

where by virtue of the small-argument representation of the modified Bessel function,

$$I_n(x) \underset{x\to 0}{\approx} \left(\frac{x}{2}\right)^n \frac{1}{n!}$$

we have ignored all but the $n = 0$ term. As one can expect, there is no temperature or recoil effect for rigidly fixed absorbers so each resonance is described by its "natural" line shape. It is to be noted that this result is not equivalent to the zero-temperature limit because in that limit the zero-point vibration effect is still to be taken into account. For the latter case $\sigma_c(K)$ is given by Eq. 4.47 multiplied by the factor $\exp(-\eta K^2)$.

The form of the cross section in Eq. 4.46 is not convenient for examining the weak binding limit. For this purpose we return to a consideration of $\langle\ \rangle$. Upon the introduction of a delta function and its integral representation, Eq. 4.42 becomes

$$\begin{aligned}
\langle\ \rangle &= \int_{-\infty}^{\infty} \mathrm{d}\varrho\, \delta(\varrho - E_k^A + E_{k''}^{A+1}) \sum_{kk''} P_k \frac{|\langle k|\, \mathrm{e}^{-i\mathbf{K}\cdot\mathbf{R}}\, |k''\rangle|^2}{(\mathscr{E}_\alpha - E_K - \varrho)^2 + (\Gamma_\alpha/2)^2} \\
&= \frac{1}{2\pi} \int_{-\infty}^{\infty} \mathrm{d}t \sum_{kk''} P_k \langle k|\, \mathrm{e}^{itH}\, \mathrm{e}^{-i\mathbf{K}\cdot\mathbf{R}}\, \mathrm{e}^{-itH}\, |k''\rangle \times \\
&\quad \times \langle k''|\, \mathrm{e}^{i\mathbf{K}\cdot\mathbf{R}}\, |k\rangle \int_{-\infty}^{\infty} \mathrm{d}\varrho\, \frac{\mathrm{e}^{-it\varrho}}{(\mathscr{E}_\alpha - E_K - \varrho)^2 + (\Gamma_\alpha/2)^2} \\
&= \frac{1}{\Gamma} \int_{-\infty}^{\infty} \mathrm{d}t\, \langle \mathrm{e}^{-i\mathbf{K}\cdot\mathbf{R}(t)}\, \mathrm{e}^{i\mathbf{K}\cdot\mathbf{R}} \rangle_T\, \mathrm{e}^{-it(\mathscr{E}_\alpha - E_K) - \Gamma_\alpha |t|/2}
\end{aligned} \tag{4.48}$$

where

$$\langle e^{-i\mathbf{K}\cdot\mathbf{R}(t)} e^{i\mathbf{K}\cdot\mathbf{R}} \rangle_T = \sum_k P_k \langle k | e^{-i\mathbf{K}\cdot\mathbf{R}(t)} e^{i\mathbf{K}\cdot\mathbf{R}} | k \rangle = e^{g(t)} \quad (4.49)$$

As in Eq. 4.45 it can be shown that

$$g(t) = \eta K^2[\coth z(\cos \hbar\omega t - 1) - i \sin \hbar\omega t] \quad (4.50)$$

If now $g(t)$ is expanded in a power series and terms up to second order in t are retained,

$$g(t) \approx -iE_R t - E_R \bar{E} t^2/2 \quad (4.51)$$

with

$$\bar{E} = \hbar\omega \coth z = \hbar\omega[\langle n \rangle_T + \tfrac{1}{2}] \quad (4.52)$$

where $\langle n \rangle_T$ is the average oscillator level. It is seen that $\bar{E}$ represents the average energy of a vibrational degree of freedom including zero-point vibration. For the truncated series to be valid it is necessary that the integrand in Eq. 4.48 make negligible contribution whenever $\hbar\omega t \gtrapprox 1$. If this is to be the case, the sum $\Gamma/2 + \mathcal{R}\mathrm{e}g(t)$ must be large and positive; in other words we require

$$\Gamma + \bar{E}(E_R/\hbar\omega) \gg 2\hbar\omega \quad (4.53)$$

which is the condition of weak binding.

The approximate form of $g(t)$ given by Eq. 4.51 allows us to express the crystal cross section in terms of the $\psi(\xi, x)$ integral introduced for the free-atom cross section.[16,19] We observe that Eq. 4.48 can now be written as

$$\langle\,\rangle \approx \frac{2}{\Gamma^2} \int_{-\infty}^{\infty} dy\, e^{-|y|} e^{-(2iy/\Gamma)(\mathscr{E} - E_K + E_R) - ((\Delta/\Gamma)y)^2} \quad (4.54)$$

where $\Delta^2 = 4E_R\bar{E}$ and the α subscript is suppressed. Again making use of the delta function and its integral representation, we have

$$e^{-|y|} = \frac{1}{2\pi} \int_{-\infty}^{\infty} ds \int_{-\infty}^{\infty} dz\, e^{isz - |z| - isy} = \frac{1}{\pi} \int_{-\infty}^{\infty} ds \frac{e^{-isy}}{1 + s^2} \quad (4.55)$$

With the help of this expression the y integration in Eq. 4.54 yields

$$\langle\,\rangle = \frac{4}{\Gamma^2} \psi(\xi, x) \quad (4.56)$$

with $\xi = \Gamma/\Delta$ and $x = 2(E_K - \mathscr{E} - E_R)/\Gamma$. The quantity x is defined here somewhat differently from that given in Eq. 4.15; the difference is of order $(m/M)^2$. Eq. 4.56 shows that the resonance line shape in crystals for the case of weak binding is the same as that in gases, but by comparing the two Δ's the crystal line shape is seen to correspond to an effective temperature of $\bar{E}/k_B$. This result was first obtained by Lamb.[17]

Elastic Scattering

An elastic scattering is an interaction in which the number of all kinds of particles, translational kinetic energy, and momentum are conserved. Evidently, in the previous chapter because of our preoccupation with the specifically nuclear aspects of neutron-nuclear reactions, we treated this notion rather casually. There, we implied that a process was elastic if the initial and final "internal" nuclear states were the same. Literally, such an implication is never justified. Practically, it is justified in the present discussion if the target nuclei are aggregated in an ideal, monatomic gas, since we have ignored all neutron-electron interactions and hence electronic excitation of atoms. But also practically, it is not justified here if the nuclei experience appreciable binding as they do in molecules, crystals, and liquids. In fact, in these latter instances, an elastic collision is one in which both the "internal" and "external" initial and final nuclear states are the same. We shall henceforth adhere to this more careful interpretation.

The reduction of R_s as given by Eq. 4.16 and 4.17 to give cross sections describing potential, resonant, and interference scatterings can be carried out without approximation. The resonant cross section will be examined in the limit of short lifetime of the compound nucleus, which, as will be seen, is quite similar to the above weak binding limit. Various aspects of the potential cross section are of interest, and these will be discussed and used to predict the behavior of the total cross section.

The reduction of R_s involves the evaluation of matrix elements of the type given in Eq. 4.44. We rewrite (4.17) as

$$\Theta = \sum_{l=1}^{4} \Theta_l \tag{4.57}$$

$$\Theta_1 = \sum_{ll'} U_l U_{l'} \langle k|e^{-i\mathbf{Q}\cdot\mathbf{R}_{l'}}|k'\rangle \langle k'|e^{i\mathbf{Q}\cdot\mathbf{R}_l}|k\rangle \tag{4.58}$$

$$\Theta_2 = \Theta_3^* = -i \sum_{ll'\alpha} U_l\, U_{0\alpha}^{Nl}\, U_{\alpha 0}^{Nl} \int_{-\infty}^{\infty} \mathrm{d}s\, \mathrm{e}^{-is(\Delta_\alpha - (i/2)\Gamma_\alpha)} \times$$

$$\times \langle k' | \mathrm{e}^{i\mathbf{Q}\cdot\mathbf{R}_{l'}} | k \rangle \langle k | \mathrm{e}^{-i\mathbf{K}\cdot\mathbf{R}_l} \mathrm{e}^{-i\mathbf{K}'\cdot\mathbf{R}_l(-s)} | k' \rangle \tag{4.59}$$

$$\Theta_4 = \sum_{ll'\alpha} U_{0\alpha}^{Nl}\, U_{\alpha 0}^{Nl}\, (U_{0\alpha}^{Nl'}\, U_{\alpha 0}^{Nl'})^* \int \mathrm{d}s\, \mathrm{d}s'\, \mathrm{e}^{is(\Delta_\alpha + (i/2)\Gamma_\alpha)}\, \mathrm{e}^{-is'(\Delta_\alpha - (i/2)\Gamma_\alpha)} \times$$

$$\times \langle k' | \mathrm{e}^{i\mathbf{K}'\cdot\mathbf{R}_l(-s)} \mathrm{e}^{i\mathbf{K}\cdot\mathbf{R}_l} | k \rangle \langle k | \mathrm{e}^{-i\mathbf{K}\cdot\mathbf{R}_{l'}} \mathrm{e}^{-i\mathbf{K}'\cdot\mathbf{R}_{l'}(-s')} | k' \rangle \tag{4.60}$$

where $\Delta_\alpha = \mathscr{E}_\alpha - E_K$ and in writing Eq. 4.60 it is assumed that the resonances do not overlap. Note that now there will be contributions from terms with $l \neq l'$, these terms lead to diffraction effects which for crystals cannot be ignored. The calculation of Θ_i is tedious but proceeds in a completely straightforward manner. We shall display only the results in the form of a differential cross section in final neutron energy and scattering angle, i.e.

$$\sigma_s(E' \to E, \theta) = \sigma_s^P(E' \to E, \theta) + \sigma_s^i(E' \to E, \theta) + \sigma_s^r(E' \to E, \theta) \tag{4.61}$$

where the superscripts denote the potential, interference, and resonant contributions, and where

$$\sigma_s^P(E' \to E, \theta) = \chi^2 \left(\frac{E}{E'}\right)^{1/2} \mathrm{e}^{-DQ^2} \left\{ \sum_{n=-\infty}^{\infty} \delta(\Delta E - n\hbar\omega)\, I_n(PQ^2)\, \mathrm{e}^{-nz} + \delta(\Delta E) \frac{1}{N_A} \sum_{ll'}{}' \mathrm{e}^{-i\mathbf{Q}\cdot(\mathbf{x}_l - \mathbf{x}_{l'})} \right\} \tag{4.62}$$

$$\sigma_s^i(E' \to E, \theta) = \chi \left(\frac{E}{E'}\right)^{1/2} \mathrm{e}^{-DQ^2} \sum_\alpha \lambda \Gamma_\alpha^{(N)} \times$$

$$\times \left\{ \mathrm{e}^{-D(K^2+K'^2)/2} \sum_{n_1 n_2 n_3} (-)^{n_1}\, \delta[\Delta E - (n_1 + n_2)\hbar\omega]\, \mathrm{e}^{-(n_1+n_2+n_3)Z} \times \right.$$

$$\times \frac{I_{n_1}(P\mathbf{Q}\cdot\mathbf{K}')\, I_{n_2}(P\mathbf{Q}\cdot\mathbf{K})\, I_{n_3}(P\mathbf{K}\cdot\mathbf{K}')\, [\mathscr{E}_\alpha - E' - (n_1 + n_3)\hbar\omega]}{[\mathscr{E}_\alpha - E' - (n_1 + n_3)\hbar\omega]^2 + (\Gamma_\alpha/2)^2}$$

$$\left. + \delta(\Delta E)\, \mathrm{e}^{-DK^2} \frac{1}{N_A} \sum_{ll'}{}' \mathrm{e}^{i\mathbf{Q}\cdot(\mathbf{x}_l - \mathbf{x}_{l'})} \sum_n \frac{I_n(P\mathbf{K}\cdot\mathbf{K}')\, (\mathscr{E}_\alpha - E' - n\hbar\omega)\, \mathrm{e}^{-nZ}}{(\mathscr{E}_\alpha - E' - n\hbar\omega)^2 + (\Gamma_\alpha/2)^2} \right\} \tag{4.63}$$

$$\sigma_s^r(E' \to E, \theta) = \left(\frac{E}{E'}\right)^{1/2} \sum_\alpha \left[\frac{\lambda \Gamma_\alpha^{(N)}}{2}\right]^2 \times$$

$$\times \Bigg\{ e^{-D(K^2+K'^2)} \sum_{n_1 \ldots n_6} (-)^{n_5+n_6} \delta(\Delta E - [n_1 + n_2 + n_5 + n_6]\hbar\omega) \times$$

$$\times\ I_{n_1}(PK^2)\, I_{n_2}(PK'^2) \times \left[\prod_{i=3}^{6} I_{n_i}(P\mathbf{K}\cdot\mathbf{K}')\, e^{-n_i Z}\right] e^{-(n_1+n_2)Z} \times$$

$$\times \frac{[\mathscr{E}_\alpha - E' - (n_2 + n_4 + n_6)\hbar\omega]\,[\mathscr{E}_\alpha - E' - (n_2 + n_3 + n_5)\hbar\omega] + (\Gamma_\alpha/2)^2}{\{[\mathscr{E}_\alpha - E' - (n_2 + n_4 + n_6)\hbar\omega]^2 + (\Gamma_\alpha/2)^2\} \times \{[\mathscr{E}_\alpha - E' - (n_2 + n_3 + n_5)\hbar\omega]^2 + (\Gamma_\alpha/2)^2\}}$$

$$+\ \delta(\Delta E)\, e^{-2DK^2} \frac{1}{N_A} \sum_{ll'}{}' e^{i\mathbf{Q}\cdot(\mathbf{x}_l - \mathbf{x}_{l'})} \times$$

$$\times \sum_{n_1 n_2} \frac{I_{n_1}(P\mathbf{K}\cdot\mathbf{K}')\, I_{n_2}(P\mathbf{K}\cdot\mathbf{K}')\,[(\mathscr{E}_\alpha - E' - n_1\hbar\omega)(\mathscr{E}_\alpha - E' - n_2\hbar\omega) + (\Gamma_\alpha/2)^2]\, e^{-(n_1+n_2)Z}}{[(\mathscr{E}_\alpha - E' - n_1\hbar\omega)^2 + (\Gamma_\alpha/2)^2]\,[(\mathscr{E}_\alpha - E' - n_2\hbar\omega)^2 + (\Gamma_\alpha/2)^2]} \Bigg\} \tag{4.64}$$

The equilibrium position of the lth nucleus (atom) is denoted by $\mathbf{x}_l$. In each cross section the direct terms ($l = l'$) have been separated from the "interference" terms ($l \neq l'$), the latter involve no energy transfer and therefore contribute only to elastic processes.* A discussion of the potential cross section will be deferred until later. From Eqs. 4.63 and 4.64 it can be observed that the influences of chemical binding upon resonance phenomenon are quite complicated and that interpretation of these results appears feasible only in the limiting cases.† The dependence of σ_s^r upon the scattering angle appears solely in the argument of the modified Bessel functions and in $\mathbf{Q}$. In the event of 90° scattering Eq. 4.64 is considerably simplified since all the n_i's except n_1 and n_2 are zero and the corresponding I_{n_i}'s are to be replaced by unity.

The exponential $\exp[-D(K^2 + K'^2)]$, in the resonant scattering cross section is known as the Lamb-Mössbauer factor. This factor provides an attenuation of any resonance process that is influenced by

---

* This is a direct consequence of the assumption of independent vibrations. In a more realistic model which treats the atomic motions as coupled oscillations there will then be both elastic and inelastic "interference" or diffraction effects.

† See, for example, the discussion of elastic resonant scattering by Trammell.[20] Analogous expressions employing more realistic models of crystals have been obtained by E. Wissler (unpublished).

temperature and lattice binding through the parameter D. On the other hand, the corresponding exponential in the potential scattering cross section is $\exp[-D(\mathbf{K} - \mathbf{K}')^2]$, which is the familiar Debye-Waller factor. The attenuation of direct processes is therefore sensitive, in addition, to the angular correlation between initial and final neutron momenta. This comparison is interesting since it tends to suggest that if the lifetime of the compound nucleus is very short the attenuation factor of a resonance process can conceivably be expressed in an angular dependent form. We shall now investigate Eq. 4.64 in such a short lifetime limit.

As in the case of the weak binding limit in radiative capture, a "time-dependent" representation of σ_s^r is more convenient for the present purposes. If we consider only the direct terms of Eq. 4.64, we have

$$\sigma_s^r(E' \to E, \theta) = \left(\frac{E}{E'}\right)^{1/2} \sum_\alpha \left[\frac{\lambda \Gamma_\alpha^{(N)}}{2}\right]^2 \times$$

$$\times \frac{1}{2\pi} \int_{-\infty}^{\infty} \mathrm{d}t\, \mathrm{e}^{-it\Delta E} \int_0^\infty \mathrm{d}s\, \mathrm{d}s'\, \mathrm{e}^{-(s+s')\Gamma_\alpha/2}\, \mathrm{e}^{-i(s-s')(\mathscr{E}_\alpha - E')}\, \mathrm{e}^{\mu(ss't)} \tag{4.65}$$

where

$$\mu(ss't) = \eta\{K^2 g(t) + K'^2 g^*(s - s' - t)$$
$$+ \mathbf{K} \cdot \mathbf{K}'[g(s') + g^*(s) - g(s' + t) - g^*(s - t)]\} \tag{4.66}$$

with $g(t)$ given by Eq. 4.50. For very large Γ the contribution to the s and s' integrals will come mainly from $s, s' \lessapprox \Gamma^{-1}$. This suggests that the terms in μ which depend only upon s or s' may be represented by truncated power series. Retaining only the first two terms we obtain

$$\mu(ss't) \approx -DQ^2 + i\hbar\omega\eta(\mathbf{K} \cdot \mathbf{K}')(s - s') + \mu'(ss't) \tag{4.67}$$

$$\mu'(ss't) = \frac{\eta\alpha_\pm}{2}\, \mathrm{e}^{\pm i\hbar\omega t}\, [K^2 + K'^2\, \mathrm{e}^{\pm i\hbar\omega(s-s')} - \mathbf{K} \cdot \mathbf{K}'(\mathrm{e}^{\pm i\hbar\omega s} - \mathrm{e}^{\mp i\hbar\omega s'})] \tag{4.68}$$

where $\alpha_\pm = \coth Z \pm 1$. In Eq. 4.68 the double sign denotes a sum of two terms corresponding to upper and lower signs respectively. The terms neglected in the above approximation are of order $(\hbar\omega s)^2$ and higher, so a condition for Eq. 4.67 to be applicable can be stated as $\Gamma \gg \hbar\omega$. Thus if the lattice binding is small compared to the resonance width the attenuation factor in σ_s^r is also effectively given by the Debye-Waller factor.*

The resonant-scattering cross section which one obtains by using the approximate form of μ is very similar to Eq. 4.64. We will not exhibit this result, but instead if we introduce a further approximation by writing

$$e^{\mu'(ss't)} \approx 1 + \mu'(ss't) \tag{4.69}$$

we would obtain

$$\begin{aligned}
\sigma_s^r(E' \to E, \theta) \approx \left(\frac{E}{E'}\right)^{1/2} e^{-DQ^2} \sum_\alpha \left[\frac{\lambda \Gamma_\alpha^{(N)}}{2}\right]^2 \times \\
\times \Bigg\{ \frac{\delta(\Delta E)}{(\mathscr{E}_\alpha - E' + \zeta)^2 + (\Gamma_\alpha/2)^2} \\
+ \frac{\alpha_\pm}{2}\, \delta(\Delta E \pm \hbar\omega) \Bigg[\frac{E/\hbar\omega}{(\mathscr{E}_\alpha - E' + \zeta)^2 + (\Gamma_\alpha/2)^2} \\
+ \frac{E'/\hbar\omega}{(\mathscr{E}_\alpha - E_{K'} + \zeta \pm \hbar\omega)^2 + (\Gamma_\alpha/2)^2} \\
- \frac{\tau/\hbar\omega}{(\mathscr{E}_\alpha - E' + \zeta \pm \hbar\omega - (i/2)\Gamma_\alpha)(\mathscr{E}_\alpha - E' + \zeta + (i/2)\Gamma_\alpha)} \\
- \frac{\tau/\hbar\omega}{(\mathscr{E}_\alpha - E' + \zeta - (i/2)\Gamma_\alpha)(\mathscr{E}_\alpha - E' + \zeta \pm \hbar\omega + (i/2)\Gamma_\alpha)} \Bigg] \Bigg\}
\end{aligned} \tag{4.70}$$

with $\zeta = \hbar\eta(\mathbf{K} \cdot \mathbf{K}')$. The first term in Eq. 4.70 represents the contribution from elastic scattering, and, in the remaining terms, upper and lower signs denote inelastic events in which the neutron loses or gains energy by an amount $\hbar\omega$ (one-phonon processes). Higher-order inelasticity has been neglected by virtue of our expansion in Eq. 4.69.†

In the remaining part of this chapter we shall restrict our attention to potential scattering only.‡ The cross section given in Eq. 4.62 is seen to

* If we interpret the compound nucleus lifetime as the interaction time, the above condition implies that the collision time be short compared to the characteristic vibration period in the lattice. This conclusion is in general agreement with Trammell.[20]

† This is somewhat similar to the "time" expansion first introduced by Wick.[21]

‡ For potential scattering of neutrons by crystals the reader should see the excellent review by Kothari and Singwi;[22] for the time-dependent representation approach see Sjolander;[23] a number of fundamental aspects of the general theory have been reviewed in detail by Yip, Osborn, and Kikuchi.[24]

contain χ simply as a multiplicative factor. For a system with nuclear spin I it will be necessary to consider both terms in (3.75). If furthermore we consider the system as an isotopic mixture with random distribution, then by carrying out the appropriate spin and isotopic averages[22,24] we find that the quantity χ^2 multiplying the direct and interference terms should be replaced by a_1^2 and a_2^2 respectively, where

$$a_1^2 = \left\langle \frac{I+1}{2I+1} a_+^2 \right\rangle + \left\langle \frac{I}{2I+1} a_-^2 \right\rangle$$

$$a_2^2 = \left\langle \frac{I+1}{2I+1} a_+^2 + \frac{I}{2I+1} a_-^2 \right\rangle \tag{4.71}$$

and the symbol $\langle\ \rangle$ here denotes isotopic average. The quantities a_+ and a_- characterize the interactions in which neutron and nuclear spin orientations are parallel and antiparallel, and are defined by

$$a_+ = \chi_0 + \frac{I}{2} \chi_1$$

$$a_- = \chi_0 - \frac{I+1}{2} \chi_1 \tag{4.72}$$

where

$$\chi_{0,1} = \frac{mg_i}{2\pi\hbar^2} \int d^3R\, v_{0,1}(R)\, e^{-i\mathbf{Q}\cdot\mathbf{R}} \tag{4.73}$$

In the special case of neutron-proton scattering a_+ and a_- would correspond to the conventional triplet and singlet scattering lengths, although in the present treatment they are functions of the momentum transfer. With the above modification the differential cross section for potential scattering* can be given in more general form,

$$\sigma_s(E' \to E, \theta) = \frac{1}{2\pi N\hbar} \left(\frac{E}{E'}\right)^{1/2} \int_0^\infty dt\, e^{-i\Delta Et/\hbar} \times$$

$$\times \left[a_1^2 \sum_l \Lambda_{ll}(\mathbf{Q}, t) + a_2^2 \sum_{ll'}{}' \Lambda_{ll'}(\mathbf{Q}, t)\right] \tag{4.74}$$

where

$$\Lambda_{ll'}(\mathbf{Q}, t) = \sum_k P_k \langle k|\, e^{i\mathbf{Q}\cdot\mathbf{R}_l(t)}\, e^{-i\mathbf{Q}\cdot\mathbf{R}_{l'}}\, |k\rangle \tag{4.75}$$

It is sometimes conventional to speak of σ_s in terms of its coherent and

* Henceforth we suppress the superscript p.

incoherent parts. Thus if we introduce coherent and incoherent scattering lengths as

$$a_{\text{coh}}^2 = a_2^2 \quad a_{\text{inc}}^2 = a_1^2 - a_2^2 \tag{4.76}$$

the cross section becomes

$$\sigma_s(E' \to E, \theta) = \frac{1}{2\pi N\hbar}\left(\frac{E}{E'}\right)^{1/2}\int_0^\infty \mathrm{d}t\, \mathrm{e}^{-i\Delta Et/\hbar} \times$$
$$\times \left[a_{\text{inc}}^2 \sum_l \Lambda_{ll}(\mathbf{Q}, t) + a_{\text{coh}}^2 \sum_{ll'} \Lambda_{ll'}(\mathbf{Q}, t)\right] \tag{4.77}$$

For the Einstein crystal this is not a particularly convenient representation so we shall continue to discuss the contributions from direct and interference scatterings separately.

Thus far it has been possible to calculate $\Lambda(\mathbf{Q}, t)$ directly because for the simple systems under consideration, the exact eigenstates $|k\rangle$ are known. For more complicated dynamic systems such as liquids, this approach is still straightforward but now the calculation depends upon less satisfying models for explicit forms of the wave function. There exists, however, an alternative and equivalent procedure for formulating the general scattering problem. In this approach the cross section is expressed in terms of a space- and time-dependent function which describes the dynamical properties of the scattering system,[25] so that the approximation in describing a complicated system then enters into the determination of this function. This function is defined as

$$G(\mathbf{r}, t) = (2\pi)^{-3}\int \mathrm{d}^3Q\, \mathrm{e}^{-i\mathbf{Q}\cdot\mathbf{r}}\, \frac{1}{N}\sum_{ll'} \Lambda_{ll'}(\mathbf{Q}, t) \tag{4.78}$$

and similar quantities for the $l = l$ and $l \neq l'$ terms only are denoted as $G_s(\mathbf{r}, t)$ and $G_d(\mathbf{r}, t)$. The cross section is then expressed as a four-dimensional Fourier transform

$$\sigma(E' \to E, \theta) = \frac{1}{2\pi}\left(\frac{E}{E'}\right)^{1/2}[a_{\text{inc}}^2\, S_{\text{inc}}(\mathbf{Q}, \Delta E) + a_{\text{coh}}^2\, S_{\text{coh}}(\mathbf{Q}, \Delta E)] \tag{4.79}$$

$$\hbar S_{\text{coh}}(\mathbf{Q}, \Delta E) = \int \mathrm{d}t\, \mathrm{d}^3r\, G(\mathbf{r}, t)\, \mathrm{e}^{i(\mathbf{Q}\cdot\mathbf{r} - \Delta Et/\hbar)} \tag{4.80}$$

and S_{inc} is obtained by replacing G with G_s. The function S is called the scattering law, and is a quantity in terms of which the scattering data can be analyzed and presented for use in the transport equation.[26,27,28]

The function, $G(\mathbf{r}, t)$, was introduced by van Hove[25] as a natural time-dependent generalization of the familiar pair distribution function $g(\mathbf{r})$ which describes the average density distribution as seen from a given particle in the system.[29] Aside from neutron scattering, $G(\mathbf{r}, t)$ is in fact a quantity of general interest in the statistical theory of many-body systems.[18,30] From the reality of S one has

$$G^*(\mathbf{r}, t) = G(-\mathbf{r}, -t) \tag{4.81}$$

The fact that G is in general complex implies that it cannot be interpreted as an observable. As suggested by van Hove, under classical conditions or more specifically when $\mathbf{R}(t)$ commutes with $\mathbf{R}$, G gives the probability that given a particle at the origin at $t = 0$ there will be a particle at $\mathbf{r}$ and t. A number of attempts have been made to develop a theory of slow-neutron scattering by liquids on the basis of such an interpretation.[31]*

We now return to more detailed consideration of neutron scattering by an Einstein crystal. From Eq. 4.74 we can write the cross section as

$$\begin{aligned}\sigma_s(E' \to E, \theta) &= \delta(\Delta E)\, e^{-DQ^2}[a_1^2\, I_0(PQ^2) - a_2^2] \\ &\quad + \delta(\Delta E)\, a_2^2\, e^{-DQ^2} \frac{1}{N}\left|\sum_l e^{i\mathbf{Q}\cdot\mathbf{x}_l}\right|^2 \\ &\quad + \left(\frac{E}{E'}\right)^{1/2} a_1^2\, e^{-DQ^2} \sum_{\substack{n=-\infty \\ n\neq 0}}^{\infty} \delta(\Delta E - n\hbar\omega)\, I_n(PQ^2)\, e^{-nZ}\end{aligned} \tag{4.82}$$

The elastic contributions are exhibited in two separate terms. The second term contains the interference factor

$$\frac{1}{N}\left|\sum_l e^{i\mathbf{Q}\cdot\mathbf{x}_l}\right|^2$$

where now l extends over all the scatterers in the spatial cell and N is their total number. For a cell of characteristic length $L \approx 10^{-4}$ cm, N is of order 10^{12} so that this factor gives the well-known Bragg condition for elastic interference scattering in the usual way. As a result of the assumption of uncorrelated vibrations, diffraction effects are seen

* For a discussion of the classical limit of the cross section see Aamodt, Case, Rosenbaum and Zweifel, *Phys. Rev.*, **126**: 1165 (1962). A discussion of the classical limit of $G(\mathbf{r}, t)$ has been given by Rosenbaum.[28]

to be purely elastic. This will not be the case if we employ a model that describes the nuclear motions as coupled oscillations.[24]

The $n \neq 0$ terms in σ_s constitute the inelastic portion of the cross section and these give rise to a set of equally-spaced lines in the spectrum corresponding to different phonon excitations. This structure is in marked contrast to the smooth distribution predicted by the gas result in Section A. Since the vibrational states are stationary in the harmonic approximation (infinite phonon lifetime) all lines have zero width.* It can be observed that so long as energy conservation is satisfied any inelastic process may occur. At $T = 0$ the neutron cannot gain any energy because $\exp(-nZ)$, interpretable as a measure of the probability of finding the oscillator in the nth eigenstate, vanishes.

The exponential factor $\exp(-DQ^2)$ in Eq. 4.82 is the quantum analogue of the Debye-Waller factor originally derived in X-ray diffraction to account for the effects of thermal motions of the scattering system. It attenuates all processes, particularly at high temperature or small Z; the effect does not vanish entirely at $T = 0$ because of zero-point motions of the scatterer. For very small Z the asymptotic form of the modified Bessel function

$$I_n(x) \underset{x \to \infty}{\approx} (2\pi x)^{-1/2} e^x$$

becomes applicable, the exponential part of which then cancels the Debye-Waller factor. Obviously the same situation holds for large Q^2 so we see that interference effects will be negligible in the region of high momentum transfer.†

Since $\sigma_s(E' \to E, \theta)$ is the differential cross section in energy and angle, the total potential scattering cross section $\sigma_s(E')$ is obtained upon integrating Eq. 4.82 over Ω and E. Because $\sigma_s(E')$ enters directly as a parameter in the transport equation, it is of some interest to examine its behavior on the basis of Eq. 4.82. The macroscopic system under consideration is in general not a single crystal, so the cross section should be averaged over crystal orientations. This aspect, however, is not relevant to our discussion, and therefore we will ignore it along

* For discussions of finite phonon lifetime in neutron scattering see Maradudin and Fein,[32] and Akcasu.[33]

† In a general theory which admits elastic as well as inelastic interference scattering the present remark applies only to the elastic portion which, however, usually provides the dominant diffraction effect.

with spin and isotope effects. At very low neutron energies ($E' \lessapprox 10^{-3}$ eV) the cross section predicts no appreciable elastic processes because $I_0(x)$ is essentially unity ($a_1^2 = a_2^2 = \chi^2$) and the wavelength is sufficiently long that the Bragg interference condition cannot be satisfied at any scattering angle. Also in this region ($E' < \hbar\omega$) the neutrons cannot lose energy, so the only permissible process is that by which the neutrons gain energy. The cross section therefore varies as $1/v$, and generally increases with temperature. As the incident energy is raised, elastic processes begin to contribute. A significant increase occurs when the Bragg condition which allows the largest wavelength is just satisfied. At still higher energies the interference term begins to be attenuated by the Debye-Waller factor, and, while the cross section will continue to exhibit sharp jumps as additional sets of crystal planes give rise to interference scattering, the over-all oscillatory behavior is damped. For sufficiently fast neutrons ($E' \gtrapprox 1$ eV) the dominant process is inelastic scattering in which the neutrons lose energy. Here each scatterer can be treated as a free particle so that the result in Section A is applicable. In fact, in the weak-binding limit one can show that

$$\sigma_s(E') \to 4\pi[M\chi/(M + m)]^2$$

The above remarks are illustrated in Fig. 4.1 which is in general agreement with observations for such scatterers as graphite, beryllium, and lead.[34]

All the discussions in this chapter have been concerned with monatomic systems and hence the center-of-mass degrees of freedom of the nucleus. However, in polyatomic systems, the neutron can excite all the degrees of freedom of the molecule so that internal molecular degrees of freedom also have to be considered. The intermediate scattering function $\Lambda(\mathbf{Q}, t)$ can be written as a product of two functions, one depending on center-of-mass translations and the other on the internal molecular motions. If rotation-vibration coupling is ignored, Λ can be further decomposed so that the effects of translation, rotation and vibration may be considered separately. From the standpoint of analyzing a particular experiment, it is important to treat the rotations properly since their energies are of the same order as those of translations. The presence of rotational transitions can therefore complicate any interpretation of the scattering data with regard to intermolecular forces.

The method of calculation presented in this section can be used to treat the normal modes of internuclear vibration. The influence of molecular rotations has been investigated mainly in neutron scattering by gases.[9,35] The cross section of a free rotator can be obtained rigorously,[36,37] but the application of the formalism is rather involved.[38] On the other hand, in systems where appreciable orientation-dependent

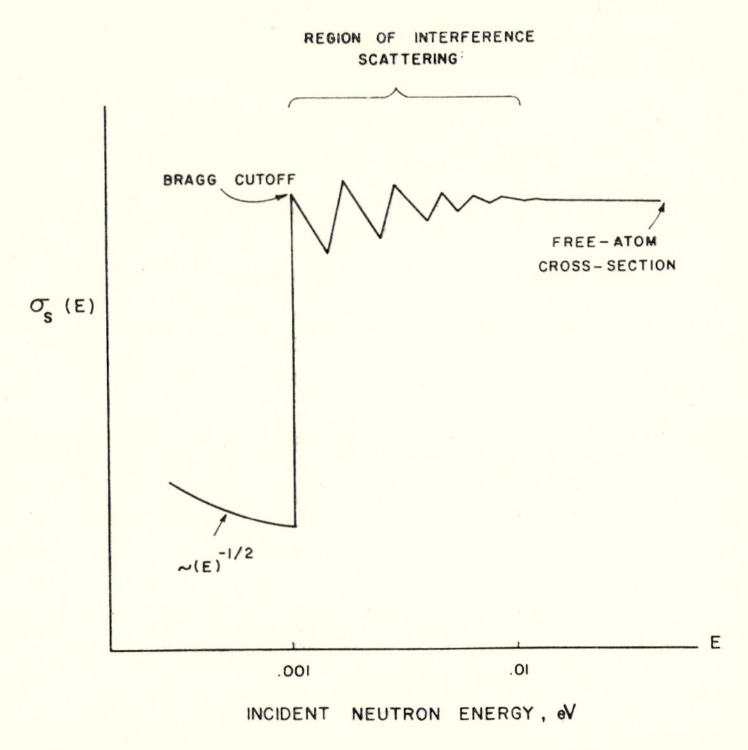

Fig. 4.1. Qualitative behaviour of total potential scattering cross section.

forces exist, rotational motions will likely become hindered. An interesting example is water where experiments have revealed prominent modes of hindered rotation. This type of motion is still not completely understood, although attempts to describe its effects in neutron scattering[39,40] are probably sufficiently accurate for thermalization calculations.

The Thermal Average

In the preceding cross-section calculations it was necessary to evaluate averages of matrix elements of the form

$$S_{ll'} = \sum_n P_n \langle n| e^{itH/\hbar} e^{-i\mathbf{K}\cdot\mathbf{R}_l} e^{-itH/\hbar} e^{i\mathbf{K}\cdot\mathbf{R}_{l'}} |n\rangle \tag{4.83}$$

for an Einstein crystal.[9] In this notation P_n is the probability that initially the crystal is in a state specified by the eigenstate $|n\rangle$, H is the crystal Hamiltonian, $\mathbf{K}$ is a momentum vector and R_l is the instantaneous position of the lth nucleus in the crystal. If equilibrium position $\mathbf{x}_l$ is introduced then

$$S_{ll'} = W_{ll'} e^{i\mathbf{K}\cdot(\mathbf{x}_{l'}-\mathbf{x}_l)} \tag{4.84}$$

$$W_{ll'} = \langle e^{-i\mathbf{K}\cdot\mathbf{u}_l(t)} e^{i\mathbf{K}\cdot\mathbf{u}_{l'}}\rangle_T \tag{4.85}$$

where we have let $\mathbf{R}_l = \mathbf{x}_l + \mathbf{u}_l$ and have introduced the Heisenberg operator

$$\mathbf{u}_l(t) = e^{itH/\hbar} \mathbf{u}_l e^{-itH/\hbar} \tag{4.86}$$

Since the Hamiltonian consists of a sum of individual particle Hamiltonians, the only part of H that does not commute with $\mathbf{u}_l$ is H_l. In Eq. 4.85 the symbol $\langle Q\rangle_T$ denotes an appropriate average of the expectation value of the operator Q. This quantity is often called the thermal average because the crystal is assumed to be initially in a thermodynamic state. Note that $W_{ll'}$ is a function of t only if $l = l'$; this is the case of direct scattering which will be considered first.

According to the Einstein model, nuclear vibrations are isotropic, so each of three directions of motion can be treated independently of the others. The fact that each nuclear coordinate is an independent oscillator coordinate reduces the calculation to a one-dimensional problem, i.e.

$$W_{ll} = \prod_{\alpha=1}^{3} W_{ll}^{\alpha} \tag{4.87}$$

$$W_{ll}^{\alpha} = \sum_{n_\alpha} P_{n_\alpha} \langle n_\alpha| e^{iK_\alpha u_\alpha(t)} e^{-iK_\alpha u_\alpha} |n_\alpha\rangle \tag{4.88}$$

For a crystal in thermodynamic equilibrium we have

$$P_{n_\alpha} = e^{-n_\alpha \hbar\omega/k_B T} \Big(\sum_{n_\alpha} e^{-n_\alpha \hbar\omega/k_\beta T}\Big)^{-1} = e^{-2Zn_\alpha}(1 - e^{-2Z}) \tag{4.89}$$

where ω is the characteristic vibrational frequency and $Z = \hbar\omega/2k_BT$. The thermal average W_{ll}^{α} can be rewritten upon the use of an operator identity

$$e^A e^B = e^{A+B+[A,B]/2} \tag{4.90}$$

which applies whenever operators A and B commute with their commutator $[A, B]$. In our case $[u_\alpha(t), u_\alpha]$ is just a c-number so that

$$W_{ll}^{\alpha} = e^{(K_\alpha^2/2)[u_\alpha(t),\, u_\alpha]} \langle e^{iK_\alpha[u_\alpha(t)-u_\alpha]}\rangle_T \tag{4.91}$$

This expression can be further simplified according to a corollary to Bloch's theorem,

$$\langle e^x \rangle_T = e^{(1/2)\langle x^2 \rangle_T} \tag{4.92}$$

where x is a multiple, or some linear combination, of commuting oscillator coordinates and their conjugate momenta. Thus

$$W_{ll}^{\alpha} = \exp\left\{-\frac{K_\alpha^2}{2}\left[\langle u_\alpha^2(t)\rangle_T + \langle u^2\rangle_T - 2\,\rangle u_\alpha(t)\, u_\alpha\rangle_T\right]\right\} \tag{4.93}$$

To evaluate the indicated thermal averages in Eq. 4.93 it is convenient to replace particle displacements by "creation" and "destruction" operators similar to those introduced in Chapter II. The new operators are governed by the commutation rule

$$[a_\alpha(t), a_{\alpha'}^+(t)] = [a_\alpha, a_{\alpha'}^+] = \delta_{\alpha\alpha'} \tag{4.94}$$

and have the properties that

$$a\,|n_\alpha\rangle = \sqrt{n_\alpha}\,|n_\alpha - 1\rangle \qquad a(t)\,|n_\alpha\rangle = \sqrt{n_\alpha}\, e^{-i\omega t}\,|n_\alpha - 1\rangle$$

$$a^+\,|n_\alpha\rangle = \sqrt{n_\alpha + 1}\,|n_\alpha + 1\rangle \qquad a^+(t)\,|n_\alpha\rangle = \sqrt{n_\alpha + 1}\, e^{i\omega t}\,|n_\alpha + 1\rangle \tag{4.95}$$

In terms of these operators,

$$u_\alpha(t) = \sqrt{\frac{\hbar}{2M\omega}}\,[a_\alpha^+(t) + a_\alpha(t)] \tag{4.96}$$

and similarly for u. The following thermal averages are then readily found,

$$\langle u_\alpha^2(t)\rangle_T = \langle u_\alpha^2\rangle_T = \frac{\hbar}{2M\omega}\,(2\,\langle n_\alpha\rangle_T + 1)$$

$$\langle u_\alpha(t)\, u_\alpha\rangle_T = \frac{\hbar}{2M\omega}\,[(\langle n_\alpha\rangle_T + 1)\, e^{-i\omega t} + \langle n_\alpha\rangle_T\, e^{i\omega t}] \tag{4.97}$$

where

$$\langle n_\alpha \rangle_T = \sum_{n_\alpha} n_\alpha P_{n_\alpha} = e^{-2Z}(1 - e^{-2Z})^{-1} \tag{4.98}$$

It is seen that in W_{ll}^{α} the only dependence upon α is in K_α^2. Thus the α product in Eq. 4.87 leads to a dependence only upon K^2 as one would expect, and we find

$$S_{ll} = e^{-DK^2} \sum_{n=-\infty}^{\infty} I_n(PK^2)\, e^{-n(Z-i\omega t)} \tag{4.99}$$

where

$$D = \eta \coth Z$$
$$P = \eta \operatorname{csch} Z$$
$$\eta = \hbar/2M\omega$$

and use has been made of the generating function of the modified Bessel function of the first kind,

$$e^{(y/2)(r+1/r)} = \sum_{n=-\infty}^{\infty} r^n I_n(y) \tag{4.100}$$

In a very similar manner the corresponding result for interference scattering ($l \neq l'$) is

$$S_{ll'} = e^{i\mathbf{K}\cdot(\mathbf{x}_{l'}-\mathbf{x}_l)} \langle e^{-i\mathbf{K}\cdot\mathbf{u}_l}\rangle_T \langle e^{i\mathbf{K}\cdot\mathbf{u}_{l'}}\rangle_T = e^{-DK^2} e^{i\mathbf{K}\cdot(\mathbf{x}_{l'}-\mathbf{x}_l)} \tag{4.101}$$

Eqs. 4.99 and 4.101 have been used to write Eqs. 4.45, 4.62, 4.63, and 4.64.

References

1. R.M. Pearce, *J. Nuclear Energy*, **A13**: 150 (1961); R.B. Nicholson, APDA-139 (1960).
2. M.S. Nelkin and E.R. Cohen, *Progress in Nuclear Energy*, Series I, **3**: 179 (1959); D.E. Parks, *Nucl. Sci. and Eng.*, **9**: 430 (1961); see also Proceedings of Brookhaven Conference on Neutron Thermalization, BNL-719 (1962).
3. *Inelastic Scattering of Neutrons in Solids and Liquids*, International Atomic Energy Agency, Vienna, 1961, 1963 (two volumes), 1965 (two volumes).
4. H.A. Bethe and G. Placzek, *Phys. Rev.*, **51**: 450 (1937).
5. H. Feshbach, G. Goertzel and H. Yamanchi, *Nucl. Sci. and Eng.*, **1**: 4 (1956).
6. J.E. Olhoeft, University of Michigan Technical Report 04261-3-F, July, 1962.
7. L. Dresner, *Nucl. Sci. and Eng.*, **1**: 68 (1956); G.M. Roe, KAPL-1241 (1954).
8. V.I. Sailor, BNL-257 (1953); M.E. Rose, W. Miranker, P. Leak, L. Rosenthal and J.K. Hendrickson, WAPD-SR-506, (1954) (two volumes).
9. A.C. Zemach and R.J. Glauber, *Phys. Rev.*, **101**: 118 (1956).
10. E.P. Wigner and J.E. Wilkins, AECD-2275 (1944).
11. B. Davison, *Neutron Transport Theory*, Oxford University Press, London, 1957.

12. K.M.Case, F.de Hoffmann, and G.Placzek, *Introduction to the Theory of Neutron Diffusion*, U.S. Government Printing Office, 1953.
13. K.M.Case, *Ann. Phys.* (N.Y.) **9**: 1 (1960).
14. M.Born and K.Huang, *Dynamical Theory of Crystal Lattices*, Oxford University Press, London, 1957.
15. H.Born and R.Oppenheimer, *Ann. Phys.*, **84**: 457 (1927).
16. A.W.McReynolds, M.S.Nelkin, M.N.Rosenbluth, and W.L.Whittemore, *Proceedings of the Second United Nations International Conference in the Peaceful Uses of Atomic Energy*, **16**: 297 (1958).
17. W.E.Lamb, *Phys. Rev.*, **55**: 190 (1939).
18. W.M.Visscher, *Ann. Phys.* N.Y., **9**: 194 (1960); M.S.Nelkin and D.E.Parks: *Phys. Rev.*, **119**: 1060 (1960); K.S.Singwi and A.Sjolander, *Phys. Rev.*, **120**, 1093 (1960).
19. R.L.Mossbauer, *Z. Physik*, **151**: 124 (1958); *Naturwissenschaften*, **45**: 538 (1958); *Z. Naturforsch.*, **14a**: 211 (1959); see also H.Frauenfelder, *The Mossbauer Effect*, Benjamin, New York, 1962.
20. G.Trammell, *Phys. Rev.*, **126**: 1045 (1962).
21. G.G.Wick, *Phys. Rev.*, **94**: 1228 (1954).
22. L.S.Kothari and K.S.Singwi, *Solid State Physics*, **8**: 109 (1959).
23. A.Sjolander, *Arkiv f. Fysik*, **14**: 315 (1958).
24. S.Yip, R.K.Osborn, and C.Kikuchi, "Neutron Acoustodynamics", NP-12399 (1963), issued as University of Michigan College of Engineering Industry Program Report IP-524.
25. L.van Hove, *Phys. Rev.*, **95**: 249 (1954).
26. P.Egelstaff, *Inelastic Scattering of Neutrons in Solids and Liquids*, International Atomic Energy Agency, Vienna, 1961, p.25; R.M.Brugger, IDO-16999 (1961).
27. M.Nelkin, *Inelastic Scattering of Neutrons in Solids and Liquids*, International Atomic Energy Agency, Vienna, 1961, p.3.
28. M.Rosenbaum and P.F.Zweifel, *Phys. Rev.* **137**: B 271 (1965).
29. F.Zernike and J.Prins, *Z. Physik*, **41**: 184 (1927); T.L.Hill, *Statistical Mechanics*, McGraw-Hill Book Co., Inc., New York, 1956.
30. U.Fano, *Phys. Rev.*, **103**: 1202 (1956).
31. G.H.Vineyard, *Phys. Rev.*, **110**: 999 (1958). K.S.Singwi and A.Sjolander, *Phys. Rev.*, **120**: 1093 (1960); A.Rahman, K.S.Singwi, and A.Sjolander, *Phys. Rev.*, **126**: 997 (1962).
32. A.Maradudin and A.E.Fein, *Phys. Rev.*, **128**: 2589 (1962).
33. Z.Akcasu and R.K. Osborn, *Nuovo Cimento*, **38**: 175 (1965).
34. D.J.Hughes and J.A.Harvey, *Neutron Cross Sections*, BNL-325 (1955).
35. R.G.Sachs and E.Teller, *Phys. Rev.*, **60**: 18 (1948); N.K.Pope, *Can. J. Phys.*, **30**: 597 (1952); T.J.Krieger and M.S.Nelkin, *Phys. Rev.*, **106**: 290 (1957).
36. A.Rahman, *J. Nucl. Energy*, **13**: 128 (1961).
37. S.Yip, Thesis, University of Michigan, Ann Arbor, Michigan, 1962.
38. H.L.McMurry, *Nucl. Sci. and Eng.*, **15**: 429 (1963).
39. M.Nelkin, *Phys. Rev.*, **119**: 741 (1960); for applications of this model see BNL-719 (1962).
40. S.Yip and R.K.Osborn, *Phys. Rev.*, **130**: 1860 (1963).

V

Special Topics

In this chapter we examine some aspects of two interesting but specialized and unrelated topics. The first has to do with what might be called neutron thermodynamics, i.e., the origin, nature and applicability of a certain time-independent, velocity-space distribution for neutrons and atoms which is achieved in special circumstances. This topic is specialized only when viewed in the context of the reactor. But it will also be seen to be an important part of the general subject of gas thermodynamics.

The second topic is also to be regarded as specialized only when considered as a part of reactor technology. As presented here it is the beginning of a study of higher-order particle distributions in reactors—in particular of a few relevant doublet densities. Such studies lead to a quantitative appreciation of the phenomena of correlations and fluctuations in the distributions of various kinds of particles. In this connection it is of interest to note that reactor-type systems are perhaps uniquely suited to an experimental investigation of these matters.

A. Neutron Thermodynamics

Most attempts at an analytical study of neutron distributions in reactors explicitly divide the energy range into at least two parts, in each of which the neutron densities are treated according to approximations peculiar to the range. The lower energy part of this subdivision is referred to as the thermal range—its upper limit being some few times the (kT) of the atoms in the system. The reference to thermal, however, is presumably not solely based on its demarcation being roughly tied to the mean energy of the atoms in the reactor but also to the expectation that, at least in many instances, the neutrons themselves in this energy range will be in a quasi-thermodynamic state. In some specific instances this expectation has been essentially verified experimentally,* but in most cases it is defended merely on speculative grounds. As a part of a study of the fundamentals of neutron transport theory, it seems appropriate to probe a little for the limitations on what can or cannot be asserted in this matter.

The initial approach to the subject will be in terms of a very special problem. Consider neutrons and a single kind of atoms mixed homogeneously in gas phase. Assume, however, that these distributions are not in a steady state. It is reasonable to expect that the mixture will indeed eventually achieve some sort of steady state, so the question is: What can be said of it? This, of course, is a familiar problem in the kinetic theory of gases.

For reasons that will become apparent later, it will be assumed that the only interaction between neutrons and nuclei and between atoms that needs to be considered is potential scattering. Neutron radiative capture processes could be included in the argument if the inverse gamma-neutron reaction was also considered and if kinetic equations for gamma rays were adjoined to the equations for the neutrons and the atoms. However, such a system (analogous to a chemically reactive gas) would be one which in the equilibrium state would not only be characterized by a specific velocity distribution for the particles and γ-rays but also by a specific ratio of particle densities. A study of the kinetics of such a situation might be interesting, but it is difficult to regard it as relevant. Reasons for not including resonant-elastic scattering are a bit more obscure and hence will not be discussed at this point.†

According to these remarks, the neutron balance relation as obtained from Eq. 4.28 is

$$\frac{\partial f}{\partial t} = \int d^3K'\, d^3k'\, d^3k\, A(\mathbf{K'k'}; \mathbf{Kk})\, [g(\mathbf{K})\, g_A(\mathbf{k}) f(\mathbf{K'}) f_A(\mathbf{k'}) - g(\mathbf{K'})\, g_A(\mathbf{k'}) f(\mathbf{K}) f_A(\mathbf{k})] \tag{5.1}$$

where

$$A(\mathbf{Kk}; \mathbf{K'k'}) = \frac{\hbar^2 \chi^2}{m^2}\, \delta(E_k + E_K - E_{k'} - E_{K'})\, \delta(\mathbf{k} + \mathbf{K} - \mathbf{k'} - \mathbf{K'}) = A(\mathbf{K'k'}; \mathbf{Kk}) \tag{5.2}$$

* See, for example, B. T. Taylor, AERE-N/R-1005 (1952).

† See, however, the footnote for Eq. 5.18.

In Eq. 5.1 we have retained the factors $g(\mathbf{K}) = 1 - 4\pi^3 f(\mathbf{K})$ and $g_A(\mathbf{k}) = 1 + (2\pi)^3 f_A(\mathbf{k})$. The former enters because neutrons are fermions and the latter because it has been assumed, for the sake of illustration, that the nuclei are spinless bosons. Strictly speaking, neither of these factors should be given much consideration because of the extreme unlikelihood of finding real systems degenerate with respect to either neutrons or nuclei. Nevertheless, it is correct to keep them, and the keeping occasions no difficulty. The extent of their practical significance will be discussed later on.

To proceed further, a balance relation for the nuclear distribution function is required. This could be deduced from first principles just as has been done for the neutrons earlier, but such a derivation would be repetitious. Hence, we merely note that

$$\begin{aligned}\frac{\partial f_A}{\partial t} = &\int d^3k'\, d^3k_1'\, d^3k_1\, A_1(\mathbf{k}'\mathbf{k}_1'; \mathbf{k}\mathbf{k}_1)\, [g_A(\mathbf{k})\, g_A(\mathbf{k}_1) f_A(\mathbf{k}') f_A(\mathbf{k}_1') \\ &- g_A(\mathbf{k}')\, g_A(k_1') f_A(\mathbf{k}) f_A(\mathbf{k}_1)] \\ &+ \int d^3K'\, d^3k'\, d^3K\, A(\mathbf{K}'\mathbf{k}'; \mathbf{K}\mathbf{k})\, [g(\mathbf{K})\, g_A(\mathbf{k}) f(\mathbf{K}') f_A(\mathbf{k}') \\ &- g(\mathbf{K}')\, g_A(\mathbf{k}') f(\mathbf{K}) f_A(\mathbf{k})]\end{aligned} \tag{5.3}$$

The first term on the right-hand side describes atomic collisions with A_1 being the scattering frequency appropriate to elastic collisions between neutral atoms, while the second term represents the effect on the atomic distribution due to neutron-nuclear collisions.

Now note that a sufficient condition that the neutron and nuclear distribution functions be independent of the time is the vanishing of the integrands in Eqs. 5.1 and 5.3, i.e.,

$$\frac{f(\mathbf{K}')}{g(\mathbf{K}')}\, \frac{f_A(\mathbf{k}')}{g_A(\mathbf{k}')} = \frac{f(\mathbf{K})}{g(\mathbf{K})}\, \frac{f_A(\mathbf{k})}{g_A(\mathbf{k})} \tag{5.4a}$$

and

$$\frac{f_A(\mathbf{k}')}{g_A(\mathbf{k}')}\, \frac{f_A(\mathbf{k}_1')}{g_A(\mathbf{k}_1')} = \frac{f_A(\mathbf{k})}{g_A(\mathbf{k})}\, \frac{f_A(\mathbf{k}_1)}{g_A(\mathbf{k}_1)} \tag{5.4b}$$

Since the primed and unprimed variables are essentially the pre- and post-collision momentum variables for particles experiencing elastic collisions, it follows that the logarithms of the factors in Eqs. 5.4a and 5.4b are at most linear, scalar combinations of the collisional invariants, i.e.,

$$\ln \frac{f(\mathbf{K})}{g(\mathbf{K})} = \alpha + \zeta \cdot m\hbar\mathbf{K} + \gamma E_K \tag{5.5a}$$

and

$$\ln \frac{f_A(\mathbf{k})}{g_A(\mathbf{k})} = \alpha_A + \zeta \cdot M\hbar\mathbf{k} + \gamma E_k \tag{5.5b}$$

where α, α_A, ζ, and γ are six arbitrary constants. A little examination reveals that the arbitrariness in the constant vector ζ must be interpreted as a velocity shared by all of the particles of both components of the gas and, as such, is ignorable in the present context. After some rearrangement, one finds that Eqs. 5.5a and 5.5b imply that

$$f(\mathbf{K}) = [4\pi^3(e^{\beta(\mu+E_K)} + 1)]^{-1} \tag{5.6a}$$

and

$$f_A(\mathbf{k}) = [8\pi^3(e^{\beta(\mu_A+E_k)} - 1)]^{-1} \tag{5.6b}$$

In these latter expressions, β, μ, and μ_A are again arbitrary constants; though the structure of these steady-state solutions to Eqs. 5.1 and 5.3 strongly suggests their interpretation as the thermodynamic solutions—and hence the identification of β as $(k^B T)^{-1}$ and μ and μ_A as the chemical potentials for the neutrons and nuclei respectively. To reinforce this interpretation, construct

$$\begin{aligned} S(t) = & -k_B \int d^3K[f(\mathbf{K}, t) \ln f(\mathbf{K}, t) + g(\mathbf{K}, t) \ln g(\mathbf{K}, t)] \\ & - k_B \int d^3k[f_A(\mathbf{k}, t) \ln f_A(\mathbf{k}, t) - g_A(\mathbf{k}, t) \ln g_A(\mathbf{k}, t)] \end{aligned} \tag{5.7}$$

This function is studied because, when evaluated in the time-independent state corresponding to the distributions (5.6a) and (5.6b), it is (to within a constant) the usually accepted expression for the entropy of an ideal-gas mixture of half-integral spin fermions and zero-spin bosons in the thermodynamic state.[1] It is now our purpose to show that this time-dependent generalization of the thermodynamic entropy function monotonically increases in time until it reaches a steady state which indeed turns out to be the steady state just referred to. That is, we present an H-theorem that suggests that Eqs. 5.1 and 5.3 describe an irreversible evolution in time of the distributions $f(\mathbf{K}, t)$ and $f_A(\mathbf{k}, t)$ toward the steady states given in Eqs. 5.6a and 5.6b, and that these latter distributions are to be interpreted as the thermodynamic distributions of the gas mixture.

Differentiating S we find that

$$\frac{\mathrm{d}S}{\mathrm{d}t} = -k_B \int \mathrm{d}^3K \frac{\partial f}{\partial t} \ln f - k_B \int \mathrm{d}^3k \frac{\partial f_A}{\partial t} \ln f_A \tag{5.8}$$

where use has been made of the fact that the total number of particles of a given kind in the system is constant in time. Using Eqs. 5.1 and 5.3 to eliminate the time derivatives in Eq. 5.8 and taking maximum advantage of the symmetries of the transition probabilities $A(\mathbf{K}'\mathbf{k}'; \mathbf{K}\mathbf{k})$ and $A_1(\mathbf{k}'\mathbf{k}_1'; \mathbf{k}\mathbf{k}_1)$, we find that

$$\begin{aligned}
\frac{\mathrm{d}S}{\mathrm{d}t} = &-\frac{k_B}{2}\int \mathrm{d}^3K'\,\mathrm{d}^3k'\,\mathrm{d}^3K\,\mathrm{d}^3k\; g(\mathbf{K}')\,g_A(\mathbf{k}')\,g(\mathbf{K})\,g_A(\mathbf{k})\,A(\mathbf{K}'\mathbf{k}';\mathbf{K}\mathbf{k}) \\
&\times \ln\frac{f(\mathbf{K})f_A(\mathbf{k})\,g(\mathbf{K}')\,g_A(\mathbf{k}')}{g(\mathbf{K})\,g_A(\mathbf{k})f(\mathbf{K}')f_A(\mathbf{k}')}\left[\frac{f(\mathbf{K}')f_A(\mathbf{k}')}{g(\mathbf{K}')\,g_A(\mathbf{k}')} - \frac{f(\mathbf{K})f_A(\mathbf{k})}{g(\mathbf{K})\,g_A(\mathbf{k})}\right] \\
&-\frac{k_B}{4}\int \mathrm{d}^3k'\,\mathrm{d}^3k_1'\,\mathrm{d}^3k\,\mathrm{d}^3k_1\; g_A(\mathbf{k}')\,g_A(k_1')\,g_A(\mathbf{k})\,g_A(\mathbf{k}_1)\,\times \\
&\times A_1(\mathbf{k}'\mathbf{k}_1';\mathbf{k}\mathbf{k}_1)\ln\frac{f_A(\mathbf{k})f_A(\mathbf{k}_1)\,g_A(\mathbf{k}')\,g_A(\mathbf{k}_1')}{g_A(\mathbf{k})\,g_A(\mathbf{k}_1)f_A(\mathbf{k}')f_A(\mathbf{k}_1')}\,\times \\
&\times\left[\frac{f_A(\mathbf{k}')f_A(\mathbf{k}_1')}{g_A(\mathbf{k}')\,g_A(\mathbf{k}_1')} - \frac{f_A(\mathbf{k})f_A(\mathbf{k}_1)}{g_A(\mathbf{k})\,g_A(\mathbf{k}_1)}\right] \geq 0
\end{aligned} \tag{5.9}$$

Thus the entropy function increases in time until the distribution functions satisfy the conditions (5.4a) and (5.4b), at which time S becomes maximum and stationary. In consequence, we shall henceforth interpret the functions (5.6a) and (5.6b) as the thermodynamic distributions for neutrons and nuclei respectively in the "ideal gas" system.

However, before these distributions can be useful to us, some estimate of the parameters μ and μ_A must be made. This is accomplished by the usual normalization requirement that the various particle densities represent a definite number of particles per cm^3. Application of the requirement leads to the observation that the factors $\exp(\beta\mu)$ and $\exp(\beta\mu_A)$ are exceedingly large, except for most unlikely conditions of high density and/or low temperature and/or small mass particles. (Conditions met, for example, by the gas-like conduction electrons in some metals at room temperature, by nearly zero-temperature gases—or liquids of He^4 and He^3, and by electron-proton gases in the cores of stars.) Consequently, for reasonable reactor conditions, we may approximate

$$f(\mathbf{K}) \approx e^{-\beta E_K}/4\pi^3\, e^{\beta\mu} \quad \text{and} \quad f_A(\mathbf{k}) \approx e^{-\beta E_k}/8\pi^3\, e^{\beta\mu_A} \tag{5.10}$$

which, when properly normalized, are simply the usual Maxwell-Boltzmann distributions for classical gases. We shall regard them as so approximated for the rest of the present discussion. It should be recalled that we have so regarded the one for the neutrons in the preceding chapters also.

The above discussion provides a fairly satisfying demonstration of the plausibility of the assertion that the solutions (5.6a) and (5.6b) (or more practically (5.10)) represent the thermodynamic distributions for the neutrons and nuclei in gas phase (assuming no sources or sinks and elastic scattering only). However, it is a bit disturbing that the demonstration was presented in so restricted a context. After all, most reactors so far have been constructed in the solid or liquid phase. Furthermore most nuclear environments interact with neutrons in many other ways than elastic scattering. Many of these interactions, such as radiative capture, fission, and nuclear inelastic scattering for example, are true preventatives of the realization of the above thermodynamic states—at least under realistic conditions. But it is anticipated—and has been suggested experimentally—that the above thermal neutron distributions will be realized in other states of matter than gas phase. Thus we present an argument or two more or less germane to the point in an effort to reinforce that anticipation.

First we note that Eq. 5.1 may be rewritten as (bearing in mind the above assumption of nondegenerate systems)

$$\frac{\partial f}{\partial t} = \int d^3K'[f(\mathbf{K}')\,\mathscr{F}(\mathbf{K}' \to \mathbf{K}) - f(\mathbf{K})\,\mathscr{F}(\mathbf{K} \to \mathbf{K}')] \tag{5.11}$$

where evidently

$$\mathscr{F}(\mathbf{K}' \to \mathbf{K}) = \int d^3k'\, d^3k\, A(\mathbf{K}'\mathbf{k}'; \mathbf{K}\mathbf{k})\, f_A(\mathbf{k}') \tag{5.12a}$$

$$\mathscr{F}(\mathbf{K} \to \mathbf{K}') = \int d^3k'\, d^3k\, A(\mathbf{K}'\mathbf{k}'; \mathbf{K}\mathbf{k})\, f_A(\mathbf{k}) \tag{5.12b}$$

With the present phrasing of the equation for the neutron distribution, a sufficient condition for a steady state becomes

$$f(\mathbf{K}')\,\mathscr{F}(\mathbf{K}' \to \mathbf{K}) = f(\mathbf{K})\,\mathscr{F}(\mathbf{K} \to \mathbf{K}') \tag{5.13}$$

The scattering kernel, $\mathscr{F}$, is essentially a momentum transfer cross section times the speed of the incident neutron. If we demand that this steady state be characterized by a Maxwellian neutron distribution, we find—after a few manipulations to extract from $\mathscr{F}$ the energy transfer cross section—that

$$E'\,\mathrm{e}^{-\beta E'}\,\sigma(E' \to E) = E\,\mathrm{e}^{-\beta E}\,\sigma(E \to E') \tag{5.14}$$

That is, if the steady state is to be a thermal one for the neutrons, then the effective energy transfer cross section (which of course is presumed to incorporate an appropriate thermal distribution for the scatterers) must satisfy a detailed balance condition, Eq. 5.14.* It is noteworthy that the effective cross section for scatterers in the crystalline phase does indeed satisfy this condition as is evidenced in Eq. 4.62. Thus it is suggested that the equilibrium distribution of neutrons in crystals will also be Maxwellian.

In a second attempt to give some force to this suggestion, we consider an H-theorem for the density matrix itself. Again, it is not so much a theorem as a plausibility argument. But when phrased in terms of the density matrix rather than the singlet densities it seems to represent a significant generalization of the above discussion to arbitrary scattering systems.

Recalling Eq. 2.56, we have

$$\frac{\partial D_{nn}}{\partial t} = \sum_{n'} W_{nn'}\,(D_{n'n'} - D_{nn}) \tag{5.15}$$

Again define an entropy function by

$$S = -k_B \sum_{n} D_{nn} \ln D_{nn} \tag{5.16}$$

If the transition probability, W, has certain symmetries, it is easily demonstrated that

$$\frac{\mathrm{d}S}{\mathrm{d}t} \geq 0 \tag{5.17}$$

The *necessary* symmetry required of W in order that Eq. 5.17 hold is

* See the discussions of Hurwitz, Nelkin and Habetler, reference 2, Appendix A.

probably not known, but it is certainly sufficient that*

$$W_{nn'} = W_{n'n} \tag{5.18}$$

Actually it is not difficult to show that Eq. 5.17 holds under the weaker symmetry requirement[4]

$$\sum_{n \neq n'} W_{n'n} D_{nn} = \sum_{n \neq n'} W_{nn'} D_{nn} \tag{5.19}$$

However, as we have seen, most of the useful representations of W for the description of neutron-nuclear reactions in the energy range germane to thermodynamic considerations actually satisfy Eq. 5.18. Thus we will spend no effort here to explore the implications of weaker requirements.

The equality in Eq. 5.17 obtains if and only if $D_{n'n'} = D_{nn}$ for all states $|n\rangle$ and $|n'\rangle$ for which $W_{nn'}$ does not itself vanish. Recalling that $W_{nn'}$ is nonzero only if $E_{n'} = E_n$, it seems evident that the time derivative of the entropy will vanish whenever the density operator assumes the form of a functional of the energy, H, i.e.,

$$D \rightarrow D(H)$$

An argument suggesting a choice of a particular functional proceeds as follows. Consider a system consisting of two weakly interacting systems. The Hamiltonian will be of the form

$$H = H^A + H^B + H^{AB} \tag{5.20}$$

---

* See Heitler, reference 3, Appendix 5. Note that from Eq. 3.34 we have

$$R_{nn'} = \frac{2\pi}{\hbar} \left| V_{nn'} - \sum_{n'' \neq n, n'} \frac{V_{nn''} V_{nn''n'}}{A_{nn''} + iB_{n''}} \right|^2$$

with $A_{nn''}$ and $B_{n''}$ real, and $V^*_{nn'} = V_{n'n}$ (V Hermitian). Thus $W_{nn'}$ is symmetric if $B = 0$ or $V_{nn'} V_{n'n''} V_{n''n}$ is real. In the case of either a direct or a pure resonance event as in the cases of potential scattering and radiative capture, condition (5.18) holds to the order of the present calculations. However, when resonant scattering is included the symmetry of the corresponding transition matrix depends upon properties of the nuclear matrix elements, U^N, which have not been discussed. By assuming

$$U^N_{\alpha''0}(\mathbf{K}s)\, U^N_{0\alpha''}(\mathbf{K}'s') \approx |U^N_{\alpha''0}|^2$$

and also no overlapping resonances at in Eq. 4.20, we have effectively asserted that Eq. 5.18 is valid in this case as well. Note also that symmetry exists in the purely absorbing case only if the resonances do not overlap.

Suppose now that this system is left isolated for a sufficient time. A steady state will be reached which we anticipate will be the thermodynamic state. If the interaction between the systems is sufficiently weak, we further anticipate that the distributions of particles among the states characteristic of each separate system will be essentially determined only by the nature of that system—except that each distribution will share a parameter common to both, the temperature. That is, we expect that

$$D(H^A + H^B + H^{AB}) \approx D(H^A + H^B) = D(H^A)\, D(H^B) \qquad (5.21)$$

A solution of the functional Eqs. 5.20 (neglecting H^{AB}) and 5.21 is

$$D = Z^{-1}\, \mathrm{e}^{-\beta H} \qquad (5.22)$$

where

$$Z = \mathrm{Tr}\, \mathrm{e}^{-\beta H} \qquad (5.23)$$

so that

$$\mathrm{Tr}\, D = 1 \qquad (5.24)$$

Applying these arguments to a system consisting of a neutron gas interacting with nuclei, we have for the Hamiltonian

$$H = \sum_{Ks} \frac{\hbar^2 K^2}{2m}\, a^+(\mathbf{K}, s)\, a(\mathbf{K}, s) + H_s + V$$

$$\approx \sum_{Ks} \frac{\hbar^2 K^2}{2m}\, a^+(\mathbf{K}, s)\, a(\mathbf{K}, s) + H_s \qquad (5.25)$$

The thermal density matrix then becomes in this instance,

$$D = D^n D^s, \qquad (5.26)$$

where

$$D^n = \frac{\exp\left[-\beta \sum_{Ks} \frac{\hbar^2 K^2}{2m}\, a^+(\mathbf{K}, s)\, a(\mathbf{K}, s)\right]}{\mathrm{Tr}\exp\left[-\beta \sum_{Ks} \frac{\hbar^2 K^2}{2m}\, a^+(\mathbf{K}, s)\, a(\mathbf{K}, s)\right]} \qquad (5.27a)$$

and

$$D^s = \frac{\exp\,[-\beta H^s]}{\mathrm{Tr}\exp\,[-\beta H^s]} \qquad (5.27b)$$

The neutron density corresponding to this density matrix is

$$f(\mathbf{K}) = \mathrm{Tr}\left(\sum_s a^+(\mathbf{K}, s)\, a(\mathbf{K}, s)\right) D$$
$$= \mathrm{Tr}\left(\sum_s a^+(\mathbf{K}, s)\, a(\mathbf{K}, s)\right) D^n \tag{5.28}$$

A straightforward calculation leads us again to the expression (5.6a). Thus by these arguments also we find the conventional expression for a fermion density in momentum space. However, in this case the distribution of the nuclei is not restricted to gas phase.

B. Higher-Order Neutron Densities—Particularly the Doublet Density

To this point the discussion has been exclusively devoted to singlet densities (especially neutron singlet densities) and approximate equations that describe them. Actually this devotion to singlet densities has been more apparent than real, since we have in fact slid over the matter of dealing with higher-order densities whenever confronted with them. Many times above, we have casually replaced certain averages of products by products of averages. Thus, without explicit comment, we have frequently met, and disposed of, higher-order densities by approximating them by products of singlet densities. For deducing equations to describe the singlet densities, these approximations are expected to be justified in the context in which they are introduced. That is, it is not anticipated that the interpretation of measurements of quantities determined primarily by mean values will be seriously falsified by ignoring fluctuations about the mean. However, occasionally experiments are designed for the explicit purpose of measuring—directly or indirectly—these fluctuations. It is perhaps obvious that such observations cannot be interpreted in terms of mean values (singlet densities) only. Thus, to develop a framework in which these observations can be studied as well as in which to investigate the importance of the approximations referred to above, we turn now to a brief examination of higher-order stochastic quantities. Actually we shall restrict our attention almost completely to second-order densities, although the generalization necessary for the consideration of densities of arbitrary order will be seen to be trivial in principle but tedious in practice.

We define a doublet density for neutrons by

$$F_2^{(n)}(\mathbf{X}, \mathbf{K}, \mathbf{X}', \mathbf{K}', t)$$
$$= L^{-6} \sum_{ss'} \mathrm{Tr}\, a^+(\mathbf{X}, \mathbf{K}, s)\, a(\mathbf{X}, \mathbf{K}, s)\, a^+(\mathbf{X}', \mathbf{K}', s')\, a(\mathbf{X}', \mathbf{K}', s')\, D(t)$$
$$= L^{-6}\, \mathrm{Tr}\, \varrho_1(\mathbf{X}, \mathbf{K})\, \varrho_1(\mathbf{X}', \mathbf{K}')\, D(t) \tag{5.29}$$

In the representation which diagonalizes the density operator with eigenvalue $N(\mathbf{X}, \mathbf{K})$, we find

$$F_2^{(n)}(\mathbf{X}, \mathbf{K}, \mathbf{X}', \mathbf{K}', t) = L^{-6} \sum_n N(\mathbf{X}, \mathbf{K})\, N(\mathbf{X}', \mathbf{K}')\, D_{nn}(t) \tag{5.30}$$

In earlier sections we have consistently approximated averages of products like the above by products of averages; which, if done here, would lead to the statement that

$$F_2^{(n)}(\mathbf{X}, \mathbf{K}, \mathbf{X}', \mathbf{K}', t) \approx F_1^{(n)}(\mathbf{X}, \mathbf{K}, t)\, F_1^{(n)}(\mathbf{X}', \mathbf{K}', t) \tag{5.31}$$

As we have already mentioned, it is not anticipated that the error introduced by approximations like Eq. 5.31 into the descriptions of the singlet densities themselves is important. However, the interpretation of experiments which, in one fashion or another, are designed to measure the difference between the left- and right-hand sides of Eq. 5.31 surely requires a more elaborate treatment of $F_2^{(n)}$. Thus we shall briefly sketch the deduction of a transport equation for the doublet density defined in Eq. 5.30.

Recalling Eq. 2.43 and the discussion leading up to it, we find that

$$\frac{\partial F_2^{(n)}}{\partial t} - L^{-6}\, \mathrm{Tr}\, \frac{i}{\hbar} [T', \varrho_1\varrho_1']\, D(t)$$
$$\approx \sum_{nn'} [(\varrho_1\varrho_1')_{n'n'} - (\varrho_1\varrho_1')_{nn}]\, W_{n'n}\, D_{nn}(t) \tag{5.32}$$

Using the relation

$$[T', \varrho_1\varrho_1'] = \varrho_1[T', \varrho_1'] + [T', \varrho_1]\, \varrho_1' \tag{5.32a}$$

in connection with the arguments leading from Eq. 2.45 to Eq. 2.49 facilitates the calculation of the first approximation to the transport terms, and we obtain

$$\left[\frac{\partial}{\partial t} + \frac{\hbar K_j}{m}\frac{\partial}{\partial X_j} + \frac{\hbar K_j'}{m}\frac{\partial}{\partial X_j'}\right] F_2^{(n)}(\mathbf{X}, \mathbf{K}, \mathbf{X}', \mathbf{K}', t)$$
$$\approx L^{-6} \sum_{nn'} [N'(\mathbf{X}, \mathbf{K})\, N'(\mathbf{X}', \mathbf{K}') - N(\mathbf{X}, \mathbf{K})\, N(\mathbf{X}', \mathbf{K}')]\, W_{n'n}\, D_{nn}(t) \tag{5.33}$$

As in Chapter II, the remainder of our task here is the reduction of the interaction terms on the right-hand side of Eq. 5.33 to useful form giving due regard to all nuclear and macroscopic medium interactions which might significantly influence the distribution, $F_2^{(n)}$. Also the details of cross-section calculations proceed here essentially the same as in Chapters III and IV. Hence little of this detail need be recapitulated; and, since our interest is in illustrating how a theory of fluctuations and correlations may be systematically constructed and not in deriving working equations for the analyses of realistic cases, we shall confine our attention to the relevant aspects of scattering, radiative capture and fission only. Then here, as in Eq. 3.39, we can write the right-hand side of Eq. 5.33 as the sum of three sets of terms, i.e.,

$$\left[\frac{\partial}{\partial t} + \frac{\hbar K_j}{m}\frac{\partial}{\partial X_j} + \frac{\hbar K_j'}{m}\frac{\partial}{\partial X_j'}\right] F_2^{(n)}(\mathbf{X}, \mathbf{K}, \mathbf{X}', \mathbf{K}', t)$$

$$= L^{-6} \sum_{nn'} [N'(\mathbf{X}, \mathbf{K})\, N'(\mathbf{X}', \mathbf{K}') - N(\mathbf{X}, \mathbf{K})\, N(\mathbf{X}', \mathbf{K}')]\, W_{n'n}^{c}\, D_{nn}(t)$$

$$+ L^{-6} \sum_{nn'} [N'(\mathbf{X}, \mathbf{K})\, N'(\mathbf{X}', \mathbf{K}') - N(\mathbf{X}, \mathbf{K})\, N(\mathbf{X}', \mathbf{K}')]\, W_{n'n}^{s}\, D_{nn}(t)$$

$$+ L^{-6} \sum_{nn'} [N'(\mathbf{X}, \mathbf{K})\, N'(\mathbf{X}', \mathbf{K}') - N(\mathbf{X}, \mathbf{K})\, N(\mathbf{X}', \mathbf{K}')]\, W_{n'n}^{F}\, D_{nn}(t) \tag{5.34}$$

where $W_{n'n}^{c}$, $W_{n'n}^{s}$, and $W_{n'n}^{F}$ are those elements of the transition probability required for the description of radiative capture, scattering and fission, respectively.

Although it is not the simplest interaction to deal with here, we shall first consider the scattering terms. The point is that this interaction is sufficiently complicated to illustrate all the interesting features of the influence of binary interactions on the time rate of change of the doublet density, and at the same time simple enough to be described in some detail. On the other hand, the treatment of the capture reactions here is almost an obvious and trivial generalization of that required earlier in the discussion of the equation for the singlet density, whereas our discussion of the fission contributions to Eq. 5.34 must necessarily be confined to results only, as their derivation is quite tedious though straightforward.

To carry out the sum over final states, it is convenient to distinguish between the terms for which $\mathbf{X} \neq \mathbf{X}'$ and those for which $\mathbf{X} = \mathbf{X}'$. We thus write

$$\begin{aligned} I_s &= L^{-6} \sum_{nn'} [N'(\mathbf{X}, \mathbf{K})\, N'(\mathbf{X}', \mathbf{K}') - N(\mathbf{X}, \mathbf{K})\, N(\mathbf{X}', \mathbf{K}')]\, W^s_{n'n}\, D_{nn}(t) \\ &= (1 - \delta_{XX'}) L^{-6} \sum_{nn'} [N'(\mathbf{X}, \mathbf{K})\, N'(\mathbf{X}', \mathbf{K}') - N(\mathbf{X}, \mathbf{K})\, N(\mathbf{X}', \mathbf{K}')]\, W^s_{n'n}\, D_{nn}(t) \\ &\quad + \delta_{XX'} L^{-6} \sum_{nn'} [N'(\mathbf{X}, \mathbf{K})\, N'(\mathbf{X}', \mathbf{K}') - N(\mathbf{X}, \mathbf{K})\, N(\mathbf{X}', \mathbf{K}')]\, W^s_{n'n}\, D_{nn}(t) \end{aligned} \tag{5.35}$$

Our purposes here will be adequately served if we suppress all detail associated with cross-section calculations and treat the distributions of target nuclei as statistically decoupled from the neutron distributions. It is to be emphasized that the neutron and nuclear distributions will not be regarded as uncorrelated in all circumstances. In fact such a correlation is crucial to the interpretation of certain fluctuation experiments to be discussed later. In view of these remarks we find as in Eq. 2.67 that

$$W^s_{n'n} = \overline{w}^s_{\mathbf{K}_2 \to \mathbf{K}_1}\, N(\mathbf{X}, \mathbf{K}_2) \left[1 - \frac{N(\mathbf{X}, \mathbf{K}_1)}{2}\right] \tag{5.36}$$

where $\mathbf{X}$ designates the spatial cell in which the scattering event takes place and $\mathbf{K}_2$ and $\mathbf{K}_1$ represent the momentum of the neutron before and after the collision. The meaning of the quantity $\overline{w}^s_{\mathbf{K}_2 \to \mathbf{K}_1}$ is best described by the relations

$$\sum_{\mathbf{K}_1} \overline{w}^s_{\mathbf{K}_2 \to \mathbf{K}_1} = \frac{\hbar K_2}{m}\, \Sigma_s(K_2) = \frac{P_2}{m}\, \Sigma_s(P_2) \tag{5.37a}$$

and

$$\begin{aligned} \sum_{\mathbf{K}_1 \in \mathrm{d}^3 K_1} \overline{w}^s_{\mathbf{K}_2 \to \mathbf{K}_1} &= \frac{\hbar K_2}{m}\, \Sigma_s(K_2)\, \mathscr{F}(\mathbf{K}_2 \to \mathbf{K}_1)\, \mathrm{d}^3 K_1 \\ &= \frac{P_2}{m}\, \Sigma_s(P_2)\, \mathscr{F}(\mathbf{P}_2 \to \mathbf{P}_1)\, \mathrm{d}^3 P_1 \end{aligned} \tag{5.37b}$$

In Eqs. 5.37a and 5.37b Σ_s is the usual macroscopic cross section and $\mathscr{F}$ is the scattering frequency. Entering Eq. 5.36 into Eq. 5.35, the sums over both final and initial states may be carried out quite straightforwardly (bearing in mind, of course, that only binary collisions are to be considered). If furthermore we pass to the continuum in momentum space, we find that

$$\mathscr{I}_s = \left(\frac{2\pi\hbar}{L}\right)^6 \Bigg[\int \mathrm{d}^3P'' \, \frac{P''}{m} \, \Sigma_s(P'') \, \mathscr{F}(\mathbf{P}'' \to \mathbf{P}) \times$$

$$\times \left\{ f_2^{(n)}(\mathbf{X}, \mathbf{P}'', \mathbf{X}', \mathbf{P}') - \frac{(2\pi\hbar)^3}{2} f_3^{(n)}(\mathbf{X}, \mathbf{P}'', \mathbf{X}', \mathbf{P}', \mathbf{X}, \mathbf{P}) \right\}$$

$$+ \int \mathrm{d}^3P'' \, \frac{P''}{m} \, \Sigma_s(P'') \, \mathscr{F}(\mathbf{P}'' \to \mathbf{P}') \times$$

$$\times \left\{ f_2^{(n)}(\mathbf{X}, \mathbf{P}, \mathbf{X}', \mathbf{P}'') - \frac{(2\pi\hbar)^3}{2} f_3^{(n)}(\mathbf{X}, \mathbf{P}, \mathbf{X}', \mathbf{P}'', \mathbf{X}', \mathbf{P}') \right\} \Bigg]$$

$$- \left(\frac{2\pi\hbar}{L}\right)^6 \frac{P}{m} \, \Sigma_s(P) \times$$

$$\times \left\{ f_2^{(n)}(\mathbf{X}, \mathbf{P}, \mathbf{X}', \mathbf{P}') - \frac{(2\pi\hbar)^3}{2} f_3^{(n)}(\mathbf{X}, \mathbf{P}, \mathbf{X}', \mathbf{P}', \mathbf{X}, \mathbf{P}'') \right\}$$

$$- \left(\frac{2\pi\hbar}{L}\right)^6 \frac{P'}{m} \, \Sigma_s(P') \times$$

$$\times \left\{ f_2^{(n)}(\mathbf{X}, \mathbf{P}, \mathbf{X}', \mathbf{P}') - \frac{(2\pi\hbar)^3}{2} f_3^{(n)}(\mathbf{X}, \mathbf{P}, \mathbf{X}', \mathbf{P}', \mathbf{X}', \mathbf{P}'') \right\}$$

$$+ \left(\frac{2\pi\hbar}{L}\right)^6 \frac{\delta_{XX'}}{L^3} \Bigg[\delta(\mathbf{P} - \mathbf{P}') \frac{P}{m} \, \Sigma_s(P) \, f_1^{(n)}(\mathbf{X}, \mathbf{P})$$

$$- (2\pi\hbar)^3 \, \delta(\mathbf{P} - \mathbf{P}') \frac{P}{m} \, \Sigma_s(P) \int \mathrm{d}^3P'' \, f_2^{(n)}(\mathbf{X}, \mathbf{P}, \mathbf{X}, \mathbf{P}'') \, \mathscr{F}(\mathbf{P} \to \mathbf{P}'')$$

$$+ \delta(\mathbf{P} - \mathbf{P}') \int \mathrm{d}^3P'' \, \frac{P''}{m} \, \Sigma_s(P'') \, \mathscr{F}(\mathbf{P}'' \to \mathbf{P}) \, f_1^{(n)}(\mathbf{X}, \mathbf{P}'')$$

$$- (2\pi\hbar)^3 \, \delta(\mathbf{P} - \mathbf{P}') \int \mathrm{d}^3P'' \, \frac{P''}{m} \, \Sigma_s(P'') \, \mathscr{F}(\mathbf{P}'' \to \mathbf{P}) \, f_2^{(n)}(\mathbf{X}, \mathbf{P}, \mathbf{X}, \mathbf{P}'')$$

$$+ \frac{P}{m} \, \Sigma_s(P) \, \mathscr{F}(\mathbf{P} \to \mathbf{P}') \left\{ 1 - \left(\frac{2\pi\hbar}{L}\right)^3 \delta(\mathbf{P} - \mathbf{P}') \right\} \times$$

$$\times \Big\{ -f_1^{(n)}(\mathbf{X}, \mathbf{P}) + (2\pi\hbar)^3 \, f_2^{(n)}(\mathbf{X}, \mathbf{P}, \mathbf{X}, \mathbf{P}) + (2\pi\hbar)^3 \, f_2^{(n)}(\mathbf{X}, \mathbf{P}, \mathbf{X}, \mathbf{P}')$$

$$+ \frac{(2\pi\hbar)^6}{2} f_3^{(n)}(\mathbf{X}, \mathbf{P}, \mathbf{X}, \mathbf{P}', \mathbf{X}, \mathbf{P}') - \frac{(2\pi\hbar)^6}{2} f_3^{(n)}(\mathbf{X}, \mathbf{P}, \mathbf{X}, \mathbf{P}, \mathbf{X}, \mathbf{P}')\Big\}$$

$$+ \frac{P'}{m} \Sigma_s(P') \mathscr{F}(\mathbf{P}' \to \mathbf{P}) \left\{1 - \left(\frac{2\pi\hbar}{L}\right)^3 \delta(\mathbf{P} - \mathbf{P}')\right\} \times$$

$$\times \Big\{ -f_1^{(n)}(\mathbf{X}, \mathbf{P}') + (2\pi\hbar)^3 f_2^{(n)}(\mathbf{X}, \mathbf{P}', \mathbf{X}, \mathbf{P}') + (2\pi\hbar)^3 f_2^{(n)}(\mathbf{X}, \mathbf{P}, \mathbf{X}, \mathbf{P}')$$

$$+ \frac{(2\pi\hbar)^6}{2} f_3^{(n)}(\mathbf{X}, \mathbf{P}, \mathbf{X}, \mathbf{P}, \mathbf{X}, \mathbf{P}') - \frac{(2\pi\hbar)^6}{2} f_3^{(n)}(\mathbf{X}, \mathbf{P}, \mathbf{X}, \mathbf{P}', \mathbf{X}, \mathbf{P}')\Big\}\Big] \tag{5.38}$$

The first four terms in this expression are strict analogues of the corresponding scattering contributions to the balance relation for the singlet density as seen in Eq. 5.1 for example. It is evident that the dependence of these terms upon the third-order densities is purely a quantum effect, and here also is of importance only for degenerate systems. The remaining terms, nonvanishing only at the points $\mathbf{X} = \mathbf{X}'$ are classical and quantum contributions to a correlation effect resulting from the fact that at $\mathbf{X} = \mathbf{X}'$ scattering can transfer neutrons between momentum cells *within* the doublet density. The factor $(2\pi\hbar/L)^6$ to which $\mathscr{I}_s$ is proportional is common to all terms in the balance relation and may be ignored throughout henceforth. In the classical limit (limit as $h \to 0$) Eq. 5.38 becomes

$$\mathscr{I}_s \to \int d^3P'' \frac{P''}{m} \Sigma_s(P'') \mathscr{F}(\mathbf{P}'' \to \mathbf{P}) f_2^{(n)}(\mathbf{X}, \mathbf{P}'', \mathbf{X}', \mathbf{P}')$$

$$+ \int d^3P'' \frac{P''}{m} \Sigma_s(P'') \mathscr{F}(\mathbf{P}'' \to \mathbf{P}') f_2^{(n)}(\mathbf{X}, \mathbf{P}, \mathbf{X}', \mathbf{P}'')$$

$$- \frac{P}{m} \Sigma_s(P) f_2^{(n)}(\mathbf{X}, \mathbf{P}, \mathbf{X}', \mathbf{P}') - \frac{P'}{m} \Sigma_s(P') f_2^{(n)}(\mathbf{X}, \mathbf{P}, \mathbf{X}', \mathbf{P}')$$

$$+ \frac{\delta_{XX'}}{L^3} \Big[\delta(\mathbf{P} - \mathbf{P}') \frac{P}{m} \Sigma_s(P) f_1^{(n)}(\mathbf{X}, \mathbf{P})$$

$$+ \delta(\mathbf{P} - \mathbf{P}') \int d^3P'' \frac{P''}{m} \Sigma_s(P'') \mathscr{F}(\mathbf{P}'' \to \mathbf{P}) f_1^{(n)}(\mathbf{X}, \mathbf{P}'')$$

$$- \frac{P}{m} \Sigma_s(P) \mathscr{F}(\mathbf{P} \to \mathbf{P}') f_1^{(n)}(\mathbf{X}, \mathbf{P}) - \frac{P'}{m} \Sigma_s(P') \mathscr{F}(\mathbf{P}' - \mathbf{P}) f_1^{(n)}(\mathbf{X}, \mathbf{P}')\Big] \tag{5.39}$$

The remainder of our considerations in this chapter will be confined to this classical limit.

Our discussion of the "capture" contributions to Eq. 5.34 can be brief. We find, in the continuum,

$$I_c = L^{-6} \sum_{nn'} [N'(\mathbf{X}, \mathbf{K})\, N'(\mathbf{X}', \mathbf{K}') - N(\mathbf{X}, \mathbf{K})\, N(\mathbf{X}', \mathbf{K}')]\, W^c_{n'n}\, D_{nn}(t)$$

$$\mathscr{I}_c = -\frac{P}{m} \Sigma_c(P) f_2^{(n)}(\mathbf{X}, \mathbf{P}, \mathbf{X}', \mathbf{P}') - \frac{P'}{m} \Sigma_c(P') f_2^{(n)}(\mathbf{X}, \mathbf{P}, \mathbf{X}', \mathbf{P}')$$

$$+ \frac{\delta_{XX'}}{L^3} \delta(\mathbf{P} - \mathbf{P}') \frac{P}{m} \Sigma_c(P) f_1(\mathbf{X}, \mathbf{P}) \tag{5.40}$$

The structure of this result is evidently the same as that seen in Eq. 5.38 (or 5.39). That is, the first two terms are analogous to the capture terms appearing in the singlet balance relation, whereas the last is inhomogeneous and implies a capture contribution to correlation. These inhomogeneous terms are interpreted to imply correlation to the extent that their presence in the balance relation for the doublet density prevents solutions of the form

$$f_2^{(n)}(\mathbf{X}, \mathbf{P}, \mathbf{X}', \mathbf{P}') = f_1^{(n)}(\mathbf{X}, \mathbf{P}) f_1^{(n)}(\mathbf{X}', \mathbf{P}') \tag{5.41}$$

The contribution of fissions to Eq. 5.34 is calculated to be

$$I_F = L^{-6} \sum_{nn'} [N'(\mathbf{X}, \mathbf{K})\, N'(\mathbf{X}', \mathbf{K}') - N(\mathbf{X}, \mathbf{K})\, N(\mathbf{X}', \mathbf{K}')]\, W^F_{n'n}\, D_{nn}(t)$$

$$\mathscr{I}_F = -\frac{P}{m} \Sigma_F(P) f_2^{(n)}(\mathbf{X}, \mathbf{P}, \mathbf{X}', \mathbf{P}') - \frac{P'}{m} \Sigma_F(P') f_2^{(n)}(\mathbf{X}, \mathbf{P}, \mathbf{X}', \mathbf{P}')$$

$$+ \int d^3P''\, f_2^{(n)}(\mathbf{X}, \mathbf{P}'', \mathbf{X}', \mathbf{P}') \frac{P''}{m} \Sigma_F(P'') \sum_{J\alpha} \alpha\, B^J_\alpha(\mathbf{P}'', \mathbf{P})$$

$$+ \int d^3P''\, f_2^{(n)}(\mathbf{X}, \mathbf{P}, \mathbf{X}', \mathbf{P}'') \frac{P''}{m} \Sigma_F(P'') \sum_{J\alpha} \alpha\, B^J_\alpha(\mathbf{P}'', \mathbf{P}')$$

$$+ \frac{\delta_{XX'}}{L^3} \Bigg[-f_1^{(n)}(\mathbf{X}, \mathbf{P}) \frac{P}{m} \Sigma_F(P) \sum_{J\alpha} \alpha\, B^J_\alpha(\mathbf{P}, \mathbf{P}')$$

$$- f_1^{(n)}(\mathbf{X}, \mathbf{P}') \frac{P'}{m} \Sigma_F(P') \sum_{J\alpha} \alpha\, B^J_\alpha(\mathbf{P}', \mathbf{P})$$

$$+ \int d^3P'' f_1^{(n)}(\mathbf{X}, \mathbf{P}'') \frac{P''}{m} \Sigma_F(P'') \sum_{J\alpha\sigma} \alpha\, \sigma B^J_{\alpha\sigma}(\mathbf{P}'', \mathbf{P}, \mathbf{P}')$$

$$+ \delta(\mathbf{P} - \mathbf{P}') f_1^{(n)}(\mathbf{X}, \mathbf{P}) \frac{P}{m} \Sigma_F(P) \tag{5.42}$$

In this equation, we have introduced some frequency functions representative of various aspects of the distribution of neutrons produced by fission. These are:*

$B^J_\alpha(\mathbf{P}'', \mathbf{P})\, d^3P \equiv$ The probability of a fission induced by a neutron at $\mathbf{P}''$ producing exactly J neutrons, α of which have momenta in d^3P about $\mathbf{P}$, (5.43a)

$B^J_{\alpha\sigma}(\mathbf{P}''\, \mathbf{P}, \mathbf{P}')\, d^3P\, d^3P' \equiv$ The probability of a fission induced by a neutron at $\mathbf{P}''$ producing exactly J neutrons, α of which have momenta in d^3P about $\mathbf{P}$ and σ of which have momenta in d^3P' about $\mathbf{P}'$. It is to be noted that $B^J_{\alpha\sigma}(\mathbf{P}'', \mathbf{P}, \mathbf{P}) = \delta_{\alpha o}\, (\mathbf{P} - \mathbf{P}')\, B^J_\alpha(\mathbf{P}'', \mathbf{P})$. (5.43b)

Evidently, for a given value of J, α and σ take on all integral values and zero subject to the condition that their sum not exceed J.

Converting the left-hand side of Eq. 5.34 to the continuum in momentum space, and substituting Eqs. 5.39, 5.40 and 5.42 into the right-hand side of Eq. 5.34 provides us with a transport equation for the neutron doublet density. One aspect of this equation is notable and requires comment. Though the equation is inhomogeneous (terms proportional to the singlet density appear in it) it does not contain the usual inhomogeneities proportional to the triplet densities (at least not in this classical limit). The absence of such terms is a direct consequence of neglecting neutron-neutron collisions and of treating the nuclear densities as known and determined independently of the neutron distributions. As we shall see below, the latter of these simplifications will have to be discarded, at least in part, if certain fluctuation experiments are to be understood and analyzed.†

An example of problems in fluctuation analysis might be the moment analysis of the record of counts by a BF_3 neutron detector. In this in-

---

* See also Eq. 2.80 and Eq. 3.101. Note that

$$\sum_{J,\alpha} \alpha B^J_\alpha(\mathbf{P}', \mathbf{P}) = \mathscr{L}(\mathbf{P}' \to \mathbf{P})$$

† For recent work on the theory of neutron fluctuation in reactors see Pluta,[5] Matthes,[6] and Bell.[7] In all these investigations the theory was developed on the basis of an ensemble probability for the reactor and the phenomenological derivation of an equation to describe it.

stance the particles detected are the alpha particles produced in the $B(n, \text{He})$ Li reaction. The count record may be analyzed in a variety of ways. Perhaps the most straightforward is to divide the record into a large number of equal time intervals and record the number of counts per interval. These numbers may then be averaged, squared and then averaged, etc. to obtain any moment of the alpha particle accumulation that is desired. The same count record may then be redivided into time intervals of a different width and the moments recomputed. This process is repeated until the desired moments have been obtained as functions of the interval width. If, for example, the mean value of the count rate changes with time, certain rather obvious refinements of the analysis of the record must be introduced. The theory to be sketched here will deal with the first two moments obtained as indicated.

The first thing to be noted is that the actual observations have nothing directly to do with neutron distributions at all. However, it will be seen that the second moment of the alpha particle accumulation (which may in fact be interpreted as the doublet density for the alphas) is coupled to the second-order cross density for alphas and neutrons, which, in turn, is coupled to the neutron doublet density for which an equation was derived above. The neutron doublet and singlet densities are coupled also to the doublet and singlet densities for the delayed-neutron precursors. We shall ignore these latter couplings (and hence delayed neutrons) since they add great bulk to the analysis but nothing new in principle. Of course, the actual interpretation of the experiment requires their consideration.

From these remarks, it is evident that even the limited treatment envisaged here will require more equations than have been derived so far. Specifically we require balance relations for the singlet and doublet densities for the alpha particles and for the doublet alpha-neutron cross density. These densities are defined by

$$F_1^{(\alpha)}(\mathbf{X}, \mathbf{K}, t) = L^{-3} \operatorname{Tr} a_\alpha^+(\mathbf{X}, \mathbf{K})\, a_\alpha(\mathbf{X}, \mathbf{K})\, D(t) \tag{5.44a}$$

$$F_2^{(\alpha)}(\mathbf{X}, \mathbf{K}, \mathbf{X}', \mathbf{K}', t) = L^{-6} \operatorname{Tr} a_\alpha^+(\mathbf{X}, \mathbf{K})\, a_\alpha(\mathbf{X}, \mathbf{K})\, a_\alpha^+(\mathbf{X}', \mathbf{K}')\, a_\alpha(\mathbf{X}', \mathbf{K}')\, D(t) \tag{5.44b}$$

$$F_2^{(\alpha n)}(\mathbf{X}, \mathbf{K}, \mathbf{X}', \mathbf{K}', t) = L^{-6} \operatorname{Tr} a_\alpha^+(\mathbf{X}, \mathbf{K})\, a_\alpha(\mathbf{X}, \mathbf{K})\, a_n^+(\mathbf{X}', \mathbf{K}')\, a_n(\mathbf{X}', \mathbf{K}')\, D(t) \tag{5.44c}$$

where the (α, n) subscripts and/or superscripts refer to quantities appropriate to alpha particles and neutrons respectively. Equations describing these densities in the classical limit and in the momentum continuum are

$$\left(\frac{\partial}{\partial t} + \frac{P_j}{M}\frac{\partial}{\partial X_j}\right) f_1^{(\alpha)}(\mathbf{X}, \mathbf{P}, t)$$
$$= \int d^3P' \frac{P'}{m} \Sigma_D(P') \mathscr{F}^D(\mathbf{P}' \to \mathbf{P}) f_1^{(n)}(\mathbf{X}, \mathbf{P}', t) \tag{5.45a}$$

$$\left(\frac{\partial}{\partial t} + \frac{P_j}{M}\frac{\partial}{\partial X_j} + \frac{P'_j}{M}\frac{\partial}{\partial X'_j}\right) f_2^{(\alpha)}(\mathbf{X}, \mathbf{P}, \mathbf{X}', \mathbf{P}', t)$$
$$= \int d^3P'' \frac{P''}{m} \Sigma_D(P'') \mathscr{F}^D(\mathbf{P}'' \to \mathbf{P}) f_2^{(\alpha n)}(\mathbf{X}', \mathbf{P}', \mathbf{X}, \mathbf{P}'', t)$$
$$+ \int d^3P'' \frac{P''}{m} \Sigma_D(P'') \mathscr{F}^D(\mathbf{P}'' \to \mathbf{P}) f_2^{(\alpha n)}(\mathbf{X}, \mathbf{P}, \mathbf{X}', \mathbf{P}'', t)$$
$$+ \delta(\mathbf{X} - \mathbf{X}')\,\delta(\mathbf{P} - \mathbf{P}') \int d^3P'' \frac{P''}{m} \Sigma_D(P'') \mathscr{F}^D(\mathbf{P}'' \to \mathbf{P}) f_1^{n}(\mathbf{X}, \mathbf{P}'', t) \tag{5.45b}$$

$$\left(\frac{\partial}{\partial t} + \frac{P_j}{M}\frac{\partial}{\partial X_j} + \frac{P'_j}{m}\frac{\partial}{\partial X'_j} + \frac{P'}{m}\Sigma_T(P')\right) f_2^{(\alpha n)}(\mathbf{X}, \mathbf{P}, \mathbf{X}', \mathbf{P}', t)$$
$$= \int d^3P'' \frac{P''}{m} \Sigma_s(P'') \mathscr{F}(\mathbf{P}'' \to \mathbf{P}') f_2^{(\alpha n)}(\mathbf{X}, \mathbf{P}, \mathbf{X}', \mathbf{P}'', t)$$
$$+ \int d^3P'' \frac{P''}{m} \Sigma_F(P'') \sum_{J=0}^{\infty} \sum_{\alpha=0}^{J} \alpha\, B_\alpha^J(\mathbf{P}'', \mathbf{P}') f_2^{(\alpha n)}(\mathbf{X}, \mathbf{P}, \mathbf{X}', \mathbf{P}'', t)$$
$$+ \int d^3P'' \frac{P''}{m} \Sigma_D(P'') \mathscr{F}^D(\mathbf{P}'' \to \mathbf{P}) f_2^{(nn)}(\mathbf{X}', \mathbf{P}', \mathbf{X}, \mathbf{P}'', t)$$
$$- \frac{\delta_{XX'}}{L^3} \frac{P'}{m} \Sigma_D(P') \mathscr{F}^D(\mathbf{P}' \to \mathbf{P}) f_1^{(n)}(\mathbf{X}, \mathbf{P}', t) \tag{5.45c}$$

In these equations we have introduced the notation $\Sigma_D(P)$ for the macroscopic cross section for the absorption of a neutron of momentum $\mathbf{P}$ by the boron in the detector, $\Sigma_T(P) = \Sigma_s(P) + \Sigma_c(P) + \Sigma_F(P) + \Sigma_D(P)$, and $\mathscr{F}^D(\mathbf{P} \to \mathbf{P}')\, d^3P'$ for the probability that, if a neutron of momentum $\mathbf{P}$ is absorbed by a boron nucleus, an alpha particle of momentum $\mathbf{P}'$ in d^3P' will be produced.

The experiment with the alpha-particle count record referred to above consists of a measurement of the singlet and doublet density for the alpha particles defined in Eqs. 5.44a and 5.44b. Thus the interpretation of that experiment in terms of system parameters requires (at least) the solution of the coupled system of Eqs. 5.45a, 5.45b, 5.45c, 5.34 (taking into account Eqs. 5.39, 5.40 and 5.42), and the equation for the neutron singlet density discussed in the previous chapters. And even this formidable task is unrealistic since it overlooks the influence of delayed neutrons. The kinetics of these "fluctuations" appears to be sensitive to these neutrons, and consequently balance relations for the singlet and doublet densities for the delayed-neutron precursors must be considered along with those just referred to. Approximate attempts to deal realistically with these experiments have been made.[5]

To facilitate a few final remarks regarding the structure of the equations for the singlet and doublet neutron densities, we recapitulate them here. We display them this time as functions of velocity rather than momentum and keep the explicit indication of the arguments of functions to a minimum. Correspondingly the equation for the singlet density reads

$$\left(\frac{\partial}{\partial t} + V_j \frac{\partial}{\partial X_j} + R_T\right) f_1 = \int d^3V'' \, G(\mathbf{V}'' \to \mathbf{V}) f_1(V'') \tag{5.46}$$

and the equation for the doublet density is

$$\begin{aligned}
&\left[\frac{\partial}{\partial t} + V_j \frac{\partial}{\partial X_j} + V'_j \frac{\partial}{\partial X'_j} + R_T(V) + R_T(V')\right] f_2 \\
&\quad - \int d^3V'' \, G(\mathbf{V}'' \to \mathbf{V}) f_2(\mathbf{V}'', \mathbf{V}') - \int d^3V'' \, G(\mathbf{V}'' \to \mathbf{V}') f_2(\mathbf{V}, \mathbf{V}'') \\
&= \delta(\mathbf{X} - \mathbf{X}') \, [\delta(\mathbf{V} - \mathbf{V}') \, R_T(V) f_1(V) - G(\mathbf{V} \to \mathbf{V}') f_1(V) \\
&\quad - G(\mathbf{V}' \to \mathbf{V}) f_1(\mathbf{V}') + \delta(\mathbf{V} - \mathbf{V}') \int d^3V'' \, R_s(V'') \, \mathscr{F}(\mathbf{V}'' \to \mathbf{V}) f_1(V'') \\
&\quad + \int d^3V'' H(\mathbf{V}'' \,|\mathbf{V}, \mathbf{V}') f_1(V'')]
\end{aligned} \tag{5.47}$$

Here we have introduced the reaction rates, $R(V) = V\Sigma(V)$, and the frequencies (unnormalized);

$$G(\mathbf{V}'' \to \mathbf{V}) = R_s(V'') \, \mathscr{F}(\mathbf{V}'' \to \mathbf{V}) + R_F(V'') \sum_{J\alpha} \alpha \, B^J_\alpha(\mathbf{V}'', \mathbf{V}) \tag{5.48}$$

and

$$H(\mathbf{V}'' \,|\mathbf{V}, \mathbf{V}') = R_F(V'') \sum_{J\alpha\sigma} \alpha \, \sigma B^J_{\alpha\sigma}(\mathbf{V}'' \,|\mathbf{V}, \mathbf{V}') \tag{5.49}$$

We have also made the identification

$$\delta(\mathbf{X} - \mathbf{X}') = \delta_{XX'}/L^3 \tag{5.50}$$

where $\delta(\mathbf{X} - \mathbf{X}')$ is to be interpreted as a Dirac delta function if we regard the domain $\{\mathbf{X}\}$ as continuous. It is worth noting that

$$\int G(\mathbf{V}'' \to \mathbf{V})\, \mathrm{d}^3V = R_s(V'') + R_F(V'')\, \langle J \rangle'' \tag{5.51}$$

and

$$\int H(\mathbf{V}''\, |\mathbf{V}, \mathbf{V}')\, \mathrm{d}^3V \mathrm{d}^3V' = R_F(V'')\, \langle J^2 \rangle'' \tag{5.52}$$

where, for example, the symbol $\langle J \rangle''$ represents the mean number of neutrons produced in a fission induced by a neutron with speed V''. In many practical applications, the dependence of $\langle J \rangle''$ and $\langle J^2 \rangle''$ upon the energy of the fission-inducing neutron is ignorable to a good approximation.

Only rudimentary investigations of the coupled systems of equations like (5.46) and (5.47) for neutron distributions have been carried through so far.[8,9] Nevertheless preliminary study has suggested a number of interesting results which warrant some comment here. However, these results have been obtained in the context of one-speed diffusion theory. Thus a considerable amount of manipulation must be performed on the above equations before they appear in a form suitable to the present discussion. Most, if not all, of these manipulations are conventional, and therefore they will be merely sketched with a minimum of detail.

The reduction to one-speed form is accomplished by integrating the equations for the singlet and doublet densities over all energies. To do this it is first necessary to transform all functions from velocity space with points labeled by the coordinates (v_x, v_y, v_z) to energy-angle space with points labeled by (E, θ, φ) or $(E, \mathbf{\Omega})$ where $\mathbf{\Omega}(\theta, \varphi) = \mathbf{v}/v$. Then it is necessary to decide how integrals of products of the various functional parameters with distribution functions are to be dealt with. This issue is subtle and complex from the qualitative point of view and almost impossibly difficult from the quantitative point of view. However, we shall treat it with the utmost casualness here, hoping that the qualitative significance of conclusions to be drawn subsequently will not be seriously jeopardized thereby. To illustrate the point and our treatment of it, consider

$$\int_{E=0}^{\infty} \mathrm{d}E\, v\, \Sigma(v) f_1(\mathbf{x}, E, \mathbf{\Omega}, t) = \int_0^{\infty} \mathrm{d}E\, R(v) f_1(\mathbf{x}, E, \mathbf{\Omega}, t) \equiv r f_1(\mathbf{x}, \mathbf{\Omega}, t) \tag{5.53}$$

where

$$f_1(\mathbf{x}, \mathbf{\Omega}, t) \equiv \int_0^\infty \mathrm{d}E f_1(\mathbf{x}, E, \mathbf{\Omega}, t) \tag{5.54}$$

As definitions, these equations can hardly be quarreled with, but as a step in the direction of reducing Eqs. 5.46 and 5.47 to a more tractible form they are purely formal. In fact, the averaged reaction rate, r, is an unknown function of $\mathbf{x}$, $\mathbf{\Omega}$, and t in general. However, we shall treat it as a constant. Then consider

$$\iint \mathrm{d}E\, \mathrm{d}E'\, R(v) f_2(\mathbf{x}, E, \mathbf{\Omega}; \mathbf{x}', E', \mathbf{\Omega}'; t) \equiv \varrho f_2(\mathbf{x}, \mathbf{\Omega}; \mathbf{x}', \mathbf{\Omega}'; t) \tag{5.55}$$

$$\iint \mathrm{d}E\, \mathrm{d}E'\, R(v') f_2(\mathbf{x}, E, \mathbf{\Omega}; \mathbf{x}', \mathbf{\Omega}', E'; t) \equiv \varrho' f_2(\mathbf{x}, \mathbf{\Omega}; \mathbf{x}', \mathbf{\Omega}'; t) \tag{5.56}$$

and

$$f_2(\mathbf{x}, \mathbf{\Omega}; \mathbf{x}', \mathbf{\Omega}'; t) \equiv \iint \mathrm{d}E\, \mathrm{d}E' f_2(\mathbf{x}, E, \mathbf{\Omega}; \mathbf{x}', \mathbf{\Omega}', E'; t) \tag{5.57}$$

Here we have introduced two new, unknown different functions of the variables $\mathbf{x}$, $\mathbf{\Omega}$, $\mathbf{x}'$, $\mathbf{\Omega}'$, and t. But these too we will treat as constants. Furthermore we will regard them as the same constant. Still further we will consider them to be the same as the corresponding reaction rate, r, i.e., we set $\varrho = \varrho' = r$. This sort of approximating will be assumed everywhere in the following discussion. With these remarks (among others unspoken) in mind, we find after integrating (5.46) and (5.47) over all energies

$$\frac{\partial f_1}{\partial t} + v\,\mathbf{\Omega}\cdot\nabla f_1 + r_T f_1 - \frac{\langle J\rangle r_F}{4\pi}\int \mathrm{d}\Omega' f_1(\mathbf{\Omega}')$$
$$- r_s \int \mathrm{d}\Omega'\, p_s(\mathbf{\Omega}' \to \mathbf{\Omega}) f_1(\mathbf{\Omega}') = S \tag{5.58}$$

$$\frac{\partial f_2}{\partial t} + v\mathbf{\Omega}\cdot\nabla f_2 + v\mathbf{\Omega}'\cdot\nabla' f_2 + 2r_T f_2 - \frac{\langle J\rangle r_F}{4\pi}\int \mathrm{d}\Omega_1 f_2(\mathbf{\Omega}_1, \mathbf{\Omega}')$$
$$- \frac{\langle J\rangle r_F}{4\pi}\int \mathrm{d}\Omega_1 f_2(\mathbf{\Omega}, \mathbf{\Omega}_1) - r_s \int \mathrm{d}\Omega_1\, p_s(\mathbf{\Omega}_1 \to \mathbf{\Omega}) f_2(\mathbf{\Omega}_1, \mathbf{\Omega}')$$
$$- r_s \int \mathrm{d}\Omega_1\, p_s(\mathbf{\Omega}_1 \to \mathbf{\Omega}') f_2(\mathbf{\Omega}, \mathbf{\Omega}_1)$$
$$= f_1(\mathbf{x}, \mathbf{\Omega}, t)\, S(\mathbf{x}', \mathbf{\Omega}', t) + f_1(\mathbf{x}', \mathbf{\Omega}', t)\, S(\mathbf{x}, \mathbf{\Omega}, t)$$
$$+ \delta(\mathbf{x} - \mathbf{x}')\,\delta(\mathbf{\Omega} - \mathbf{\Omega}')\, S(\mathbf{x}, \mathbf{\Omega}, t) + \delta(\mathbf{x} - \mathbf{x}')\,\delta(\mathbf{\Omega} - \mathbf{\Omega}') \times$$
$$\times \left\{ r_T f_1 + \int \mathrm{d}\Omega_1 f_1(\mathbf{\Omega}_1) \left[r_s\, p_s(\mathbf{\Omega}_1 \to \mathbf{\Omega}) + \langle J^2\rangle \frac{r_F}{4\pi} \right] \right\}$$

$$- \delta(\mathbf{x} - \mathbf{x}') f_1(\mathbf{\Omega}') \left[r_s p_s(\mathbf{\Omega}' \to \mathbf{\Omega}) + \frac{\langle J \rangle r_F}{4\pi} \right]$$

$$- \delta(\mathbf{x} - \mathbf{x}') f_1(\mathbf{\Omega}) \left[r_s p_s(\mathbf{\Omega} \to \mathbf{\Omega}') + \frac{\langle J \rangle r_F}{4\pi} \right] \tag{5.59}$$

In these equations we have introduced a source of neutrons that is independent of the neutron density, i.e., a source other than fission. This was accomplished by simply assuming that our system includes nuclei that decay spontaneously with the emission of a neutron. The inclusion of such a source is desirable and important since ultimately we wish to explore fluctuations in both critical and subcritical systems. We have also introduced the symbol $p_s(\mathbf{\Omega} \to \mathbf{\Omega}')\, d\Omega'$ to represent the probability that, given that a neutron going in the direction $\mathbf{\Omega}$ is scattered, it is scattered to some direction $\mathbf{\Omega}'$ in $d\Omega'$, and v to represent the mean speed. Eq. 5.58 is recognized as the conventional one-speed transport equation for the neutron singlet density. Eq. 5.59 is the corresponding analogue for the neutron doublet density. It is important to note that the equation for the singlet is independent of the doublet and that the inhomogeneous terms in the equation for the doublet depend only on the singlet and the source. This linearity of these equations is a direct consequence of ignoring the neutron-neutron interaction and of the assumption that the atoms in our system are in distributions independent of the presence of the neutrons.

A diffusion approximation to these equations is now obtained by standard arguments. First, Eqs. 5.58 and 5.59 are integrated over all directions, $\mathbf{\Omega}$; then they are multiplied by $\mathbf{\Omega}$ and again integrated over directions of motion. Defining

$$n_1(\mathbf{x}, t) \equiv \int d\Omega\, f_1(\mathbf{x}, \mathbf{\Omega}, t) \tag{5.60}$$

$$\mathbf{\Phi}(\mathbf{x}, t) \equiv \int d\Omega\, \mathbf{\Omega} f_1(\mathbf{x}, \mathbf{\Omega}, t) \tag{5.61}$$

$$n_2(\mathbf{x}, \mathbf{x}', t) \equiv \int d\Omega\, d\Omega'\, f_2(\mathbf{x}, \mathbf{\Omega}; \mathbf{x}', \mathbf{\Omega}'; t) \tag{5.62}$$

$$\mathbf{\Phi}(\mathbf{x}, \mathbf{x}', t) \equiv \int d\Omega\, d\Omega'\, \mathbf{\Omega} f_2(\mathbf{x}, \mathbf{\Omega}; \mathbf{x}', \mathbf{\Omega}'; t) \tag{5.63}$$

$$\mathbf{\Phi}'(\mathbf{x}, \mathbf{x}', t) \equiv \int d\Omega\, d\Omega'\, \mathbf{\Omega}' f_2(\mathbf{x}, \mathbf{\Omega}; \mathbf{x}', \mathbf{\Omega}'; t) \tag{5.64}$$

one obtains

$$\frac{\partial n_1}{\partial t} + v\nabla \cdot \mathbf{\phi} + (r_A - \langle J \rangle r_F)\, n_1 = S_0 \tag{5.65a}$$

$$\frac{\partial}{\partial t} \mathbf{\phi} + v \int d\Omega\, \mathbf{\Omega}(\mathbf{\Omega} \cdot \nabla) f_1 + r_T \mathbf{\phi} - \bar{\mu} r_s \mathbf{\phi} = \mathbf{S}_1 \tag{5.65b}$$

and

$$\frac{\partial n_2}{\partial t} + v\nabla \cdot \mathbf{\Phi} + v\nabla' \cdot \mathbf{\Phi}' + 2(r_A - \langle J \rangle r_F) n_2$$

$$= n_1 S_0' + n_1' S_0 + \delta(\mathbf{x} - \mathbf{x}') [S_0 + r_c n_1 + \langle (J-1)^2 \rangle r_F n_1] \qquad (5.66a)$$

$$\frac{\partial}{\partial t} \mathbf{\Phi} + v \int d\Omega\, d\Omega'\, \mathbf{\Omega}(\mathbf{\Omega} \cdot \nabla) f_2 + v \int d\Omega\, d\Omega'\, \mathbf{\Omega}(\mathbf{\Omega}' \cdot \nabla') f_2$$

$$+ [2r_T - \langle J \rangle r_F - \mu r_s - r_s] \mathbf{\Phi}$$

$$= \boldsymbol{\phi} S_0' + n_1' \mathbf{S}_1 + \delta(\mathbf{x} - \mathbf{x}') [\mathbf{S}_1 + r_A \boldsymbol{\phi} - \langle J \rangle r_F \boldsymbol{\phi}] \qquad (5.66b)$$

$$\frac{\partial}{\partial t} \boldsymbol{\phi}' + v \int d\Omega\, d\Omega'\, \mathbf{\Omega}'(\mathbf{\Omega} \cdot \nabla) f_2 + v \int d\Omega\, d\Omega'\, \mathbf{\Omega}'(\mathbf{\Omega}' \cdot \nabla') f_2$$

$$+ [2r_T - \langle J \rangle r_F - \bar{\mu} r_s - r_s] \mathbf{\Phi}'$$

$$= n_1 \mathbf{S}_1' + S \boldsymbol{\phi}' + \delta(\mathbf{x} - \mathbf{x}') [\mathbf{S}_1 + r_A \boldsymbol{\phi} - \langle J \rangle r_F \boldsymbol{\phi}] \qquad (5.66c)$$

At this point numerous approximations are made. The nonfission source is assumed to be spherically symmetric so that $\mathbf{S}_1 = 0$.

The ambiguous terms are approximated (albeit somewhat inconsistently) according to the assumption that the velocity anisotropy of both densities is describable by a linear combination of the zeroeth and first spherical harmonics. The inconsistency, and its effects on the doublet equations and some of their solutions, is discussed in detail in reference 9. It will not be dealt with here since its removal greatly complicates the formal appearance of the various doublet equations without significantly altering their content so far as present illustrative purposes are concerned. The above equations then become

$$\frac{\partial n_1}{\partial t} + v\nabla \cdot \boldsymbol{\phi} + \alpha n_1 = S_0 \qquad (5.67a)$$

$$\frac{\partial}{\partial t} \boldsymbol{\phi} + \frac{v}{3} \nabla n_1 + r_{TR} \boldsymbol{\phi} = 0 \qquad (5.67b)$$

and

$$\frac{\partial n_2}{\partial t} + v\nabla \cdot \mathbf{\Phi} + v\nabla' \cdot \mathbf{\Phi}' + 2\alpha n_2$$

$$= n_1 S_0' + n_1' S_0 + \delta(\mathbf{x} - \mathbf{x}') [S_0 + \beta n_1] \qquad (5.68a)$$

$$\frac{\partial}{\partial t}\boldsymbol{\Phi} + \frac{v}{3}\nabla n_2 + (\alpha + r_{TR})\,\boldsymbol{\Phi} = \boldsymbol{\phi} S_0' + \delta(\mathbf{x} - \mathbf{x}')\,\alpha\boldsymbol{\phi} \tag{5.68b}$$

$$\frac{\partial}{\partial t}\boldsymbol{\Phi}' + \frac{v}{3}\nabla' n_2 + (\alpha + r_{TR})\,\boldsymbol{\Phi}' = S_0\boldsymbol{\phi}' + \delta(\mathbf{x} - \mathbf{x}')\,\alpha\boldsymbol{\phi}' \tag{5.68c}$$

Here we have introduced some new notation, i.e.,

$$\alpha = r_A - \langle J\rangle\, r_F \tag{5.69}$$

$$r_{TR} = r_T - \bar{\mu} r_s \tag{5.70}$$

and

$$\beta = r_c + \langle (J-1)^2\rangle\, r_F \tag{5.71}$$

These systems of coupled equations can be reduced to partial differential equations for n_1 and n_2 by straightforward manipulations. The resultant equations are of second order in the time for n_1 and third order in the time for n_2. The time derivatives of higher order than the first are required for the description of relatively fast transients. If attention is centered in slowly varying, or even temporally asymptotic, solutions, the time derivatives in the current Eqs. 5.67b, 5.68b, and 5.68c can be neglected, and one obtains

$$\frac{\partial n_1}{\partial t} - \frac{v^2}{3r_{TR}}\nabla^2 n_1 + \alpha n_1 = S_0 \tag{5.72}$$

and

$$\begin{aligned}\frac{\partial n_2}{\partial t} - \frac{v^2}{3(\alpha + r_{TR})}&(\nabla^2 + \nabla'^2)\, n_2 + 2\alpha n_2 \\ &= n_1 S_0' + n_1' S_0 + \delta(\mathbf{x} - \mathbf{x}')\,[S_0 + \beta n_1] \\ &\quad + \frac{v^2}{3r_{TR}(\alpha + r_{TR})}(S_0'\nabla^2 n_1 + S_0\nabla'^2 n_1') \\ &\quad + \frac{2\alpha v^2}{3r_{TR}(\alpha + r_{TR})}\,\delta(\mathbf{x} - \mathbf{x}')\,\nabla^2 n_1\end{aligned} \tag{5.73}$$

The arguments leading to Eq. 5.73 are a bit devious and some explanation is required. According to Eqs. 5.68b and 5.68c, $\boldsymbol{\Phi}$ and $\boldsymbol{\Phi}'$ depend upon terms containing Dirac delta functions. This is all right as it stands but becomes a somewhat ambiguous matter when these relations are inserted into Eq. 5.68a. We have circumvented this ambiguity by replacing the delta function by a Gaussian which is equivalent to a delta function in the limit of vanishing width. But this replacement leads to uncertainty as to whether the singlet current in the term should be evaluated at the primed or the unprimed point. In the absence of clear-cut evidence indicating a choice in this matter, we have chosen to evaluate these singlet currents as one-half of the sum of their values at the two different points. Thus, for example, we have employed the replacement

$$\alpha\delta(\mathbf{x} - \mathbf{x}')\,\phi \rightarrow \alpha\,\frac{e^{-|\mathbf{x}-\mathbf{x}'|^2/4\sigma}}{(4\pi\sigma)^{3/2}}\,\frac{1}{2}\,(\phi + \phi') \tag{5.74}$$

taking the limit $\sigma \rightarrow 0$ at a convenient point in the analysis. Eqs. 5.72 and 5.73 constitute the "diffusion description" of the neutron singlet and doublet densities.

Before considering a solution to these equations in an interesting special case, it is useful to examine a general characteristic of the neutron doublet density as described by Eq. 5.47. For this purpose it is convenient to display Eqs. 5.46 and 5.47 more compactly as

$$\frac{\partial f_1}{\partial t} + Bf_1 = S \tag{5.75}$$

and

$$\frac{\partial f_2}{\partial t} + (B + B')f_2 = Sf_1' + S'f_1 + \delta(\mathbf{x} - \mathbf{x}')\,\delta(\mathbf{v} - \mathbf{v}')\,S + \delta(\mathbf{x} - \mathbf{x}')\,\Gamma f_1 \tag{5.76}$$

Here again we have introduced the external source function, $S(\mathbf{x}, \mathbf{v}, t)$. The integration of Eqs. 5.75 and 5.76 over energy yields Eqs. 5.58 and 5.59 including source terms. The operator B we will refer to as the Boltzmann operator. Its definition is readily inferred from Eq. 5.46. Similarly the operator Γ, appearing in one of the inhomogeneous terms on the right-hand side of Eq. 5.76, is obtained by reference to Eqs. 5.47, 5.48, and 5.49. The prime on the Boltzmann operator in Eq. 5.76 implies that it operates on the primed phase point upon which $f_2(\mathbf{x}, \mathbf{v}, \mathbf{x}', \mathbf{v}', t)$ depends.

For some purposes it may be convenient to study the correlation function (or generalized variance of the distribution) defined as

$$G(\mathbf{x}, \mathbf{v}, \mathbf{x}', \mathbf{v}', t) = f_2(\mathbf{x}, \mathbf{v}, \mathbf{x}', \mathbf{v}', t) - f_1(\mathbf{x}, \mathbf{v}, t)\,f_1(\mathbf{x}', \mathbf{v}', t) \tag{5.77}$$

Straightforward exploitation of Eqs. 5.75 and 5.76 demonstrates that the correlation function satisfies the equation

$$\frac{\partial G}{\partial t} + (B + B')\, G = \delta(\mathbf{x} - \mathbf{x}')\, \delta(\mathbf{v} - \mathbf{v}')\, s + \delta(\mathbf{x} - \mathbf{x}')\, \Gamma f_1 \tag{5.78}$$

The primary reason for reopening the discussion of the general Eqs. 5.75 and 5.76, or 5.75 and 5.78, which we prefer in the present instance, is to extract a comment on the so-called critical state of a nuclear reactor. Thus we examine—in rather formal terms—a solution to the source-free equations

$$\frac{\partial f_1}{\partial t} + B f_1 = 0 \tag{5.79}$$

and

$$\frac{\partial G}{\partial t} + (B + B')\, G = \delta(\mathbf{x} - \mathbf{x}')\, \Gamma f_1 \tag{5.80}$$

Assume the existence of a complete, orthogonal (discrete and/or continuous) set of eigenfunctions and corresponding eigenvalues, which satisfy

$$B\psi_j = \lambda_j \psi_j \tag{5.81}$$

as well as appropriate no-reentrant-current boundary conditions over the surface of the reactor. Of course, the existence of such a function set has not yet been established. Nevertheless we surmise that the principal inference to be drawn from the present argument has a wide validity, even though its logical defense is not immediately accessible. Thus, proceeding purely formally, we expand

$$f_1(\mathbf{x}, \mathbf{v}, t) = \sum_j a_j(t)\, \psi_j(\mathbf{x}, \mathbf{v}) \tag{5.82}$$

where the sum over the eigenlabels "j" is a sum and/or an integration corresponding to their discrete and/or continuous distribution. We then find that

$$\frac{\mathrm{d}a_j}{\mathrm{d}t} = -\lambda_j a_j \tag{5.83}$$

so that

$$a_j(t) = a_j(0)\, \mathrm{e}^{-\lambda_j t} \tag{5.84}$$

and

$$f_1 = \sum_j a_j(0)\, \mathrm{e}^{-\lambda_j t}\, \psi_j \tag{5.85}$$

We now assume that the eigenvalues can be ordered, e.g.

$$\lambda_0 < \lambda_1 < \cdots < \lambda_j < \cdots \tag{5.86}$$

and that λ_0 is a discrete eigenvalue. Then the criticality condition is $\lambda_0 = 0$, which implies

$$\lim_{t \to \infty} f_1 \to a_0(0)\, \psi_0(\mathbf{x}, \mathbf{v}) \tag{5.87}$$

Now consider Eq. 5.80, Expand

$$G(\mathbf{x}, \mathbf{v}, \mathbf{x}', \mathbf{v}', t) = \sum_{jk} A_{jk}(t)\, \psi_j(\mathbf{x}, \mathbf{v})\, \psi_k(\mathbf{x}', \mathbf{v}') \tag{5.88}$$

and

$$\delta(\mathbf{x} - \mathbf{x}'))\, \Gamma f_1 \approx \delta(\mathbf{x} - \mathbf{x}')\, a_0(0)\, \Gamma \psi_0 = \sum_{jk} C_{jk}\, \psi_j(\mathbf{x}, \mathbf{v})\, \psi_k(\mathbf{x}', \mathbf{v}') \tag{5.89}$$

The coefficients A_{jk} then satisfy

$$\frac{\mathrm{d}A_{jk}}{\mathrm{d}t} + (\lambda_j + \lambda_k)\, A_{jk} = C_{jk} \tag{5.90}$$

For $j = k = 0$ we have

$$A_{00}(t) = A_{00}(0) + tC_{00} \tag{5.91}$$

and apparently G increases without limit as $t \to \infty$, unless C_{00} is zero which is generally not to be expected. Thus the correlation function—and hence also the neutron doublet density according to the present description—has no meaning in the critical state, if that state is realized according to (5.87).

The significance and full range of validity of this result is by no means understood at this point. Consequently we do not wish to dwell upon it here nor to struggle much with its interpretation. However, we do feel that it is important to point it out as a precautionary comment regarding attempts to study neutron doublet densities in critical systems. In particular, we ourselves are guided by it in restricting our consideration of solutions of Eq. 5.73 to those appropriate to subcritical systems only.

Now returning to Eqs. 5.72 and 5.73, we seek their steady-state solutions in regions of large systems over which the neutron singlet density and source may be regarded as space-independent. In such an instance, (5.72) and (5.73) become

$$n_1 = s/\alpha \tag{5.92}$$

and

$$-\frac{v^2}{3r_{TR}(1 + \alpha/r_{TR})}(\nabla^2 + \nabla'^2)\, n_2 + 2\alpha n_2$$

$$= an_1 S + \delta(\mathbf{x} - \mathbf{x}')\,[S + \beta n_1] \tag{5.93}$$

where we have dropped the subscript on the source symbol. Noting that α/r_{TR} is generally very small compared to unity, and making use of (5.92), we may rewrite (5.93) as

$$[-D(\nabla^2 + \nabla'^2) + 2\alpha]\, n_2 = \frac{2n_1^2}{\alpha} + \delta(\mathbf{x} - \mathbf{x}')\,(\alpha + \beta)\, n_1 \tag{5.94}$$

where we have defined

$$D = v^2/3t_{TR} \tag{5.95}$$

Because of the character of the space dependence of the inhomogeneous terms in Eq. 5.94 it is evident that n_2 depends only upon $\mathbf{R} = \mathbf{x} - \mathbf{x}'$. Consequently (5.94) further simplifies to

$$[-D\nabla_R^2 + \alpha]\, n_2(\mathbf{R}) = \frac{n_1^2}{\alpha} + \delta(\mathbf{R})\,\frac{\alpha + \beta}{2}\, n_1 \tag{5.96}$$

This equation is readily solved, yielding

$$n_2(\mathbf{R}) = n_1^2 + n_1\,\frac{\alpha + \beta}{2D}\,\frac{e^{-\varkappa R}}{4\pi R} \tag{5.97}$$

where

$$\varkappa^2 = \alpha/D \tag{5.98}$$

This is substantially the same solution that was presented in reference 8.

Defining a normalized correlation function as

$$C(R) = \frac{n_2(R) - n_1^2}{n_2^2} \tag{5.99}$$

we have

$$C(R) = \frac{\alpha + \beta}{2Dn_1}\,\frac{e^{-\varkappa R}}{4\pi R} \tag{5.100}$$

We also note that (5.98) may be written in terms of the diffusion length, $L = (1/3\,\Sigma_a\,\Sigma_{TR})^{1/2}$, and infinite medium multiplication constant, $k = \langle j\rangle\,\Sigma_f/\Sigma_a$, as

$$\varkappa^2 = \frac{1 - k}{L^2} \tag{5.101}$$

Thus we see that the correlation length is here given by $\varkappa^{-1} = L/\sqrt{1-k}$. Evidently the correlation extends over very large distances in large nearly critical systems for which $\sqrt{1-k}$ is small, as well as in non-multiplying but weakly absorbing regions for which L becomes large. However, in the latter instance, $\alpha + \beta$ becomes very small so that although the correlation is widespread it is at the same time very weak. It is to be noted that the normalized correlation varies inversely as the singlet density and hence becomes less noticeable as this density increases. This is perhaps an intuitively anticipated observation.

References

1. J.E.Mayer and M.G.Mayer, *Statistical Mechanics*, John Wiley & Sons, Inc., New York, 1940; L.D.Landau and E.M.Lifshitz, *Statistical Physics*. Addison-Wesley, Reading, Mass., 1958.
2. H.Hurwitz, M.S.Nelkin and G.J.Habetler, *Nucl. Sci. and Eng.*, **1**: 280 (1956).
3. W.Heitler, *The Quantum Theory of Radiation*, Oxford University Press, New York, 1954, third edition.
4. E.C.G.Stückelberg, *Helv. Phys. Acta*, **25**: 577 (1952).
5. P.Pluta, Thesis, The University of Michigan, Ann Arbor, Michigan, 1962.
6. W.Matthes, *Nukleonik*, Band **4**, Heft 5: 213 (1962).
7. G.I.Bell, *Ann. Phys.*, **21**: 243 (1963).
8. R.K.Osborn and S.Yip, Proceedings of the Conference on Noise Analysis, University of Florida, November, 1963.
9. R.K.Osborn and M.Natelson, *J. Nucl. Energy*, Part A/B, **19**: 619 (1965).

Index

Manufactured in Great Britain